41st Annual Meeting of the Plant Growth Regulation Society of America 2014

San Francisco, California, USA
13-17 July 2014

ISBN: 978-1-5108-2884-1

Printed from e-media with permission by:

Curran Associates, Inc.
57 Morehouse Lane
Red Hook, NY 12571

Some format issues inherent in the e-media version may also appear in this print version.

Printed by Curran Associates, Inc. (2016)

For permission requests, please contact Plant Growth and Regulation Society of America at the address below.

Plant Growth and Regulation Society of America
1018 Duke Street
Alexandria, VA 22314
USA

Phone: 703-836-4606 x116
Fax: 703-836-4607

cslone@ashs.org

Additional copies of this publication are available from:

Curran Associates, Inc.
57 Morehouse Lane
Red Hook, NY 12571 USA
Phone: 845-758-0400
Fax: 845-758-2633
Email: curran@proceedings.com
Web: www.proceedings.com

MEETING SUMMARY

41st Annual Meeting of the Plant Growth regulation Society of America

Timothy M. Spann

President, PGRSA

California Avocado Commission, 12 Mauchly, Suite L, Irvine, CA 92618 USA

e-mail: tim@pgrsa.org

The 41st Annual Meeting of the Plant Growth Regulation Society of America (PGRSA) was held in San Francisco, California, 13-17 July 2014. The PGRSA, established in 1973, is a forum for scientists from diverse disciplines to exchange ideas and information about different facets of plant growth regulation. The primary purpose of the PGRSA is to disseminate information concerning regulation of plants growth that results in safe, environmentally sound, and efficient production of food, fiber and ornamentals. The PGRSA meets at a different location in North America each year during the summer and has met jointly with other professional societies with similar interests when it is mutually beneficial.

For our 41st Annual meeting we partnered with the Japanese Society for Chemical Regulation of Plants to bring world-class plant hormone researchers to the conference. Dr. Carl Sams, 1st Vice President of the PGRSA and Program Chair for the 2014 meeting, assembled an outstanding slate of speakers discussing plant stress tolerance, postharvest physiology, natural products, disease management and many other exciting topics. The meeting also included our traditional welcome reception, poster sessions, industry updates and a post-conference tour.

HONORARY LIFETIME MEMBERS

Anson Cook

Aldo J. Crovetti

Eric A. Curry

Horace G. Cutler

Richard Dunand

C. Dave Fritz

Masayuki Katsumi (JSCRP)

Wayne Mackay

Wilhelm Rademacher

Edward F. Sullivan

Acknowledgements

The Proceedings has been published yearly since the fourth annual meeting in 1977 at no charge to the author(s). papers in this edition were received electronically and, although we have made minor changes in formatting, every effort has been made to preserve the original submissions, the content of which remains the sole responsibility of the respective author(s).

The Proceedings were compiled by:
Dr. Timothy M. Spann, 2013 to 2014 President PGRSA
Research Program Director
California Avocado Commission
12 Mauchly, Suite L
Irvine, CA 92618 USA
Tel: 949-341-1955 FAX: 949-341-1970
tim@pgrsa.org

The Proceedings are available for download free of charge at www.PGRSA.org.

Plant Growth Regulation Society of America
Ms. Cindy Slone, Executive Secretary
1018 Duke Street
Alexandria, VA 22314
Tel: 703-836-4606
FAX: 703-836-4607
cslone@ashs.org

A SPECIAL THANK YOU TO THE FOLLOWING COMPANIES FOR FINANCIAL SUPPORT OF THE MEETING

PLATINUM SPONSORS

VALENT BIOSCIENCES. CORPORATION

StollerUSA
Unleashing The Power Of Plants

SILVER SPONSORS

ohp | Partners with solutions

acadian.

fine

2013 – 2014 PGRS Steering Committee

Dr. Tim Spann (President)
California Avocado Commission
12 Mauchly, Suite L
Irvine, CA 9261
949.754.0732 (phone)
949.208.3512 (fax)
tspann@avocado.org

Mr. Dave Barcel (Past President)
OHP, Inc.
W331S4122 Saddleback Drive
Genesee, WI. 53118
262.392.3004 (phone)
262.392.3007 (fax)
dbarcel@ohp.com

Dr. Carl Sams (1st Vice President)
Austin Distinguished Professor
The University of Tennessee
Department of Plant Sciences
Rm 252 Ellington Plant Sciences Bldg.
2431 Joe Johnson Drive
Knoxville, TN 37996-4561
865.974.8818 (phone)
865.974.1947 (fax)
carlsams@utk.edu

Dr. Holly Little (2nd Vice President)
Acadian Seaplants Ltd.
2443 Fair Oaks Blvd., #67
Sacramento, CA 95825
530-219-5415 (phone)
hlittle@acadian.ca

Dr. Chris Gunter (Secretary)
Horticultural Science
North Carolina State University
230 Kilgore Hall, Box 7609
Raleigh, NC 27619-7609
919-513-2807 (phone)
919-515-2505 (fax)
chris_gunter@ncsu.edu

Dr. Gregory Peck (MAL-1)
Assistant Professor of Horticulture
Virginia Tech
Alson H. Smith, Jr. Agricultural Research and Extension Center
595 Laurel Grove Road
Winchester, VA 22602 USA
540.869.2560 (phone)
greg.peck@vt.edu

Dr. Albert Liptay (MAL-3)
Stoller Enterprises, Inc.
4001 West Sam Houston Parkway N., Suite 100
Houston, TX 77043
713.461.1493 (phone)
aliptay@stollerusa.com

Dr. Langtao Xiao (Adjunct Member)
Hunan Agricultural University
Key Lab of Phytohormones
Changsha, Hunan
CHINA 410128
86-731-84635261 (phone)
86-731-4635261 (fax)
langtaoxiao@163.com

Dr. Sonja L. Maki (Executive Director)
Biology Department
Carleton College
1 North College Street
Northfield, MN 55057
507.222.4544 (phone)
507-222-5757 (fax)
smaki@carleton.edu

Ms. Cindy Slone (Executive Secretary)
PGRSA/American Society for Horticultural Science
1018 Duke Street
Alexandria, VA 22314
703-836-4606 x113 (phone)
703-836-2024 (fax)
cslone@ashs.org

Dr. Tadao Asami (Liaison for JSCRP)
The University of Tokyo
Department of Applied Biological Chemistr

TABLE OF CONTENTS

PERCEPTION MECHANISM OF STRIGOLACTONES AND APPROACH FOR AGRICULTURAL PROBLEMS

Tadao Asami[1,2]

[1]*Graduate School of Agricultural and Life Sciences, The University of Tokyo, Tokyo 113-8657, Japan*

[2]*Corresponding author: asami@mail.ecc.u-tokyo.ac.jp*

Strigolactones (SLs) are phytohormones that inhibit shoot branching and function in the rhizospheric communication with symbiotic fungi and parasitic weeds. An α/β-hydrolase protein, DWARF14 (D14), has been recognized to be an essential component of plant SL-signaling, although its precise function remains unknown. Here we present the SL-dependent interaction of D14 with a gibberellin (GA)-signaling repressor SLR1 and a possible mechanism of phytohormone perception in D14-mediated SL signaling. D14 functions as a cleavage enzyme of SLs, and the cleavage reaction induces the interaction with SLR1. The crystal structure of D14 shows that 5-hydroxy-3-methylbutenolide (D-OH), which is a reaction product of SLs, is trapped in the catalytic cavity of D14 to form an altered surface. The D14 residues recognizing D-OH are critical for the SL-dependent D14–SLR1 interaction. These results provide new insight into crosstalk between GA and SL signaling pathways. Our model of the SL-perception by D14 explains why a wide variety of natural SLs and analogs can exert their activity as branching inhibitors; the D-ring moiety of SLs is essential for hormonal activity. Our findings would contribute to the development of novel SL analogs and SL-signaling inhibitors for unveiling further details of SL signaling and for controlling plant growth and protecting crops from parasitic weeds in order to increase crop yields.

EFFECTS OF HOMOBRASSINOLIDE APPLIED ON ALMONDS TO ENHANCE NUT SIZE AND POLLEN TUBE GROWTH

N. Bhushan Mandava[1,4], Scott Hicks[2], Carlos Sotomayor[3] and Segundo Maita[3]

[1]Repar Corporation, 8070 Georgia Avenue , Suite 209, Silver Spring, MD 20910, USA

[2]Innovative Ag Research and Consulting, LLC, Fresno, CA, USA

[3]Pontificia Universidad Catolica de Chile, Faculty of Agronomy and Forestry Engineering, Santiago, Chile

[4]Corresponding author: bhushan@reparcorp.com

This paper deals with the HBR effects on almonds in USA and Chile. In the U.S. Study, HBR was applied on almonds (var. Non-pareil) to enhance nut size and increase the numbers of nuts retained at harvest. Rates of HBR applied were 1 ppm ai or 10 ppm ai alone plus tank mix treatments of HBR plus 1 ppm ai or 10 ppm ai CPPU (forchlorfenuron). The most consistent and significantly effective individual treatment was 10 ppm ai HBR tank mixed with 1 ppm ai CPPU. There is a synergistic effect between HBR and CPPU. In Chile, we obtained results similar to those in the U.S. field trial. We conducted a parallel *in vitro* study to find out the HBR's effects on pollen germination and pollen tube growth in almonds, which would result in better fruit set and increased fruit seed weight.

A POTENTIAL OF *SLETR1-2*, A WEAK ALLELE OF TOMATO ETHYLENE RECEPTOR MUTANT AS A BREEDING MATERIAL FOR IMPROVING SHELF LIFE OF FRUITS

Syariful Mubarok[1,3], Y. Okabe[2], T. Ariizumi[2], N. Fukuda[2] and H. Ezura[2],

[1]Graduate School of Life and Environmental Sciences, University of Tsukuba, Tsukuba, Japan and Department of Agronomy, Faculty of Agriculture, Padjadjaran University, Bandung, Indonesia

[2]Graduate School of Life and Environmental Sciences, University of Tsukuba, Tsukuba, Japan

[3]Corresponding author: mubarok_hort@yahoo.com

Ethylene is a key factor for the regulation of tomato fruit shelf life and accelerates fruit quality deterioration. Modification of ethylene signaling pathway and biosynthesis allows us to improve tomato fruit shelf life. In our previous study, we have isolated tomato mutants with a mutation in the ethylene receptor gene (*ETR1*) from our Micro-Tom mutant library. According to the preliminary evaluation, *Sletr1-2* mutant showed a reduced sensitivity to ethylene and extended shelf life of fruits, expecting the potential as a breeding material. The aim of this study was to evaluate the potential of *Sletr1-2* mutation for F1 hybrid cultivars with extended fruit shelf life. *Sletr1-2* and its background (*WT-MT*) were crossed with four commercial pure line cultivars ('Aichi First', 'Ailsa Craig', 'Money Maker' and 'M82') to obtain F1 hybrid lines that would be evaluated. The F1 hybrids were cultivated using NFT hydroponic cultivation systems and evaluated for the growth and development. Compared to *WT-MT* F1 hybrid lines, *Sletr1-2* F1 hybrid lines showed extended fruit shelf life, increased fruit firmness and slightly reduced a fruit red color while it did not show significant differences in other fruit characters such as size, fresh weight and pericarp thickness and plant development. Furthermore, the sugar content of two *Sletr1-2* F1 lines was higher than those *WT-MT* F1. These results suggest a significant potential of *Sletr1-2* as a breeding material for improving shelf life of tomato fruits.

SOYBEAN (*GLYCINE MAX*) GROWTH AND POD DISTRIBUTION IN RESPONSE TO EARLY SEASON APPLICATION OF RYZUP SMARTGRASS®

Michael D. Rethwisch[1,2], Garrett Tooker[1], and Scott Willet[1]

[1]University of Nebraska-Lincoln, Butler County Extension Office, 451 N. 5th Street, David City, NE 68632, USA

[2]Corresponding author: michael.rethwisch@unl.edu

Various aspects of soybean (*Glycine max*) growth were documented in response to RyzUp SmartGrass® applied at unifoliate leaf (0.3, 0.5 oz/acre), first trifoliate leaf, or both (0.3 + 0.3 oz/acre). All RyzUp SmartGrass® treatments resulted in highly visible and significantly taller plant heights by 4 days post treatment due to longer internode lengths, with internode differences still evident at harvest. While significantly increasing height of first trifoliate leaf node above the soil surface to increase harvest efficacy, sequential applications of RyzUp SmartGrass® also resulted in a significantly more pods/plant (11.8%) at harvest, with increased numbers of pods documented on lower plant nodes. Yield increase for the sequential application was calculated to be 3-5 bushels/acre.

THERMOBIOGENETICS™ FOR THE GROWTH AND REPRODUCTION OF CROPS

Jerry H. Stoller[1,2]

[1]*Stoller USA, 4001 W Sam Houston Pkwy N, Houston, TX 77043 USA*

[2]*Corresponding author: jstoller@stollerusa.com*

A new concept will be introduced concerning the thermo-bio-genetic activities of plants. It will cover the total energy intake of the plant and the partitioning of the energy into exothermic and endothermic energy at different stages of plant growth to maximize yield. Thermo-bio-genetics is based upon the concept of the change in temperature and the change in pressure over time. It is now obvious that a plant undergoes these dynamic changes throughout its various growth stages. This thermo-bio-genetics is primarily regulated by climatic conditions.

Climatic activities have a direct effect on the expression of the genetic algorithm of a plant and how the plant responds to such. It appears that it is possible to use exogenous applications of thermo-bio-genetic regulators to make the plant increase its energy uptake and have a more perfect energy partition throughout the different stages of plant growth.

Members of the Plant Growth Regulator Society of America and guests, I want to thank each of you for the opportunity to make this presentation entitled "Thermobiogenetics™ for the Growth and Reproduction of Crops." This is a new subject on which very little has been investigated or very few bibliographies can be cited. Therefore, this presentation is basically presented on a theoretical level with a number of field experiments that will tend to substantiate this new theoretical approach to plant science.

I want to thank each of you for keeping an open mind on the subject, while realizing that the bibliography and experiments that are presented are not significant in numbers. I believe they are significant in results. For today only, please do not judge this speech as a peer review will judge it. There will be presentations following this that will go further into the metabolical activities which are a consequence of the effects of thermodynamics on the absorption of energy by a plant and the partitioning of energy within the plant at various growth stages.

The first law of thermodynamics dictates that energy can neither be created nor destroyed. It merely changes forms. In the environment it changes from ice to water and then to liquid. In reverse the thermodynamic activity can reverse this thermodynamic of change. When we become plant specific, energy is primarily converted to growth (work), which is the energy absorbing the chlorophyll from the sunlight. At the same time, the plant converts energy to solids (endothermic energy) while the energy that is absorbed by the soils and processed through the microbes that exist on the meristematic tissue of the roots.

It is the energy that is created by the chlorophyll molecules that primarily partition energy within the plant into exothermic energy. It is the energy that is absorbed through the roots that primarily converts the partitioning energy within the plant into endothermic energy.

The energy absorbed by the apical meristem tissue (produced by chlorophyll) transverses the gradient downward in a plant which follows the typical thermodynamic pathway of transferring energy from the hottest source to the coldest source. Contrarily, the energy that is directed through the meristematic tissue of the roots transverses against the Delta T gradient due to the exothermic activity of the microbes that colonize the root tissue.

The normal formula used by scientists that categorize energy is:

$$\text{Energy} = \int_{Germination}^{Maturity} \Delta T, \Delta P, dt \quad \text{[where T = Temperature; P = Pressure and dt = with Time]}$$

In other words, the algorithm that guides the DNA of any particular plant species or variety merely has an algorithm that is driven by energy. As the energy varies due to the variance of Delta T and Delta P, the amplitude of the energy may change at any moment. The algorithm of the DNA, however, determines the relative values of Delta T and Delta P at different growth stages of each plant. This is genetically determined. It is not primarily determined by Delta T and Delta P in the absolute manner. Changes in these two parameters merely change the amplitude of which the DNA algorithm is expressed. It does not change the general direction throughout future stages of growth.

SEED GERMINATION. When a seed is planted in the soil it is warmer than the soil. Therefore, the seed must express exothermic energy and pass energy to the soil to start the process of amylase synthase which then initiates cell division and cell differentiation from the germ of the seed.

From Hypocotyl Emergence to V2 Growth Stage. It is during this stage of growth that the germinating plant receives its energy from the stored energy in the seed (roots) and begins to receive energy through its chlorophyll molecules. At this stage of growth the young seedling is still primarily receiving its energy from the seed (roots) but yet starting to receive energy from the sunlight.

V2 – Vn Stage of Growth. It is during this stage of growth that the plant receives energy from both the sunlight and the soil. It is during this period that root growth is most massive in weight and volume. Evidently, at this stage of growth the energy received from the soil is extremely important in establishing the proper partition of energy in the plant to maintain adequate cell division, cell differentiation, and cell arrangement to propel new growth from the apical meristem tissue.

At the same time, it is the IAA gradient (ΔT gradient) that moved upward in a plant which has a greater partitioning of energy for endothermic purposes. This establishes a reserve of carbohydrates and protein as a latent energy source to be released during the reproductive period of any plant.

Vn – R2 Stage of Growth. It is during this period of growth that exothermic energy in a plant must prevail. In other words, the gradient of IAA from the apical meristem downwards must have greater signaling power than the gradient of the IAA moving upward. Science has shown that during the reproductive period IAA gradient downward tends to be greatly diminished.

Unfortunately, only little research has ever determined the IAA gradient moving upwards from the roots. It is during this stage of plant development that stem cells from the nucleus of every cell develop into reproductive cells as opposed to vegetative cells, which the DNA has previously determined. It is a massive shift in the algorithm of the DNA that releases energy during this stage of growth.

R2 – Rn Stage of Growth. It is during this period that the progeny (seed, fruit, or storage tissue) is cooler than the mother plant. Since energy is always transferred from the warmer source to the cooler source, energy is then transferred from the mother plant to the progeny.

This is possibly the reason why every progeny is covered by a (husk on corn, a hull, on wheat, rye, barley, rice) or askin (fruit) which keeps the progeny cooler in temperature than the mother plant during the daytime hours.

Thermodynamic Activity as Related to IAA Transport. The law of thermodynamics dictates that energy moves from the hottest source to the coolest source. Therefore, one would expect the IAA from the top of the plant moving downward to progress through heat differential that occurs in a plant tissue. Scientists have determined that IAA moves only by gravity. Yet, scientists have also determined that IAA gradient is greatly inhibited by shade. Obviously, shade must have some sort of determination of gravity. It does not.

One can only conclude that the gradient of IAA from the apical meristem tissue downward is a function of tissue temperature differential. Therefore, Delta IAA will progress along the same gradient as Delta T.

The shaded part of the plant is exposed to less wind movement and less energy transfer. Therefore, the shaded part of the plant becomes warmer than the upper part of the plant. This is why IAA gradient moving downward is inhibited after shading occurs.

A single corn plant sitting in sunlight would have multiple ears since there is nothing to inhibit IAA gradient movement downward in a plant. When the same corn plant is planted in a high population, the corn plant only has a single ear. Think about it!!

The partitioning of energy when the IAA gradient move downward is the energy that is used for exothermic processing (ΔP). This energy is used as work energy (cell division and cooling). What then determines the relative "work energy" (exothermic) partitioning of energy in a plant? It is primarily the Delta P which represents the pressure differential of moisture at the root hairs up to the stomata where the moisture is evaporating. As the Delta P increases, the amount of exothermic energy increases. It is this release of exothermic energy that cools the plant tissue. In other words, Delta P represents the air conditioning in a plant.

By realizing this Delta P that controls the exothermic energy expressed by a plant, one can then understand why a sudden warm rain after deteriorating soil moisture will cause such an exaggerated growth in any young plant. There is immediate release of exothermic energy from the endothermic energy that is then stored in a plant during the period of drought. It is under the conditions of low Delta P (saturated soil moisture and high humidity) where exothermic energy release is at a minimum. Due to the low Delta P, the plant does not express exothermic energy therefore rapid cell division, cell differentiation, and cell arrangement is not exhibited.

Since Delta P is a variable gradient, the plant does not expend energy. It conserves energy. There is little or no energy transferred to new plant cells that tend to differentiate. The mature plant cells tend to want to store their energy. They do not want to release this energy to the new developing plant cells. Therefore, new developing plant cells exhibit an energy shortage.

Under saturated moisture conditions and/or high humidity, it is the new plant cell tissue that normally becomes energy insufficient. It is the apical meristem tissue and the development of the roots. It is the flowering positions of a plant. It is the new apical leaf tissue of a plant. It is precisely these areas that become energy deficient of the areas that are subject to disease, insect, and nematode attacks.

If this hypothesis is valid, a plant should be completely resistant to any insect, disease, or nematode, if there is adequate endothermic energy provided by the storage energy within the plant so that all new plant cells are energy sufficient. The plant cells will never have exudates that will attract any insects, diseases, or nematodes.

Again, it is the IAA signaling from the plant roots upward that express endothermic energy. This can only be possible if it is a consequence of the microbes that colonize the meristematic tissue of the roots. They must become the driving force to propel the upward gradient of IAA. The upward gradient signaling is conducted by the meristematic root tissue even before one sees the consequence of open stomata or closed stomata.

All of us should consider the possibility that IAA is a majoring signaling hormone in a plant. It signals each cell and prepares each cell to accept the consequences in the change of Delta T and Delta P. The gradient of IAA does not cause the partitioning of energy. It is merely a signaling that prepares the nucleus of every plant cell to guide the cell division and cell differentiation that will result from such.

I believe that it is important for all of us to consider IAA merely as a signaling molecule. It sets the positive polarity of plant cells. The signaling of IAA, however, is dependent upon negative polarity of a plant cell. This negative polarity is provided by cytokinin. It is the ratio of IAA to cytokinin that determines the expression of energy that is impinged upon each cell nucleus.

It is then possible to understand that two different stages of plant growth the ratio of IAA to cytokinin must change to determine the growth habits of the plants. It is all a matter of polarity between plant cells and the Delta T and Delta P that is expressed by this polarity.

At this point, it is now possible to understand that all physiological properties of a plant are driven by thermodynamics. It will drive the amplitude (upward or downward) of the plant's DNA algorithm. It will not determine the absolute direction. It will merely determine the amplitude of the DNA that has been genetically been determined.

As one can further absorb that the signaling, which enables this partitioning of energy is performed primarily by IAA and complimented by cytokinin. This is necessary to prepare themselves for the change in Delta T and Delta P at any moment in time.

SUMMARY. If we completely understand the above subject, we can then exogenously apply materials to either the seed or the plant roots to increase the microbial activity (microbial species) that convert energy from the soil and impinged into the roots of plants.

We can exogenously apply material to the plant tissue that can cause the plant tissue to become cooler so that it will absorb more energy from the sunlight.

We can exogenously apply the material to the mother plant that the time that we want to release a lot of this energy into exothermic form to cause more reproductive cell division, cell differentiation, and cell arrangement during the initial stage of the plant's reproductive life.

We can now understand why we can exogenously apply materials to the mother plant to once again be able to transfer more energy to the developing progeny.

If we employ these practices, we will always be able to utilize the energy within the plant to achieve maximum productivity. We will have the maximum energy input into the plant from the sunlight and the soil. We will manipulate the plants that increase the amplitude of the partitioning of energy at each stage of plant growth. This will give us the maximum yields.

If we accomplish the above purpose, we will automatically make the plant resistant to any disease, insects, or nematodes.

It is my hope that this presentation will inspire younger scientists to investigate the Thermobiogenetic™ activities as a new course in agronomy. I believe that they will find the consequences of doing so will greatly increase the efficiency of nutrient utilization within the plant cells. This will eliminate the need for such high qualities of nutrients, which pollute the environment in obtaining maximum yields.

It is my hope that plant pathologist and entomologist study this new science to make the plant more resistant to diseases so that fewer pesticides must be used in crop production.

I want to thank all of you for so diligently listening to my presentation. I am preparing a manuscript, which will be published in the fall to give further information for your studies.

A SIMPLIFIED, PRACTICAL, HORMONE BALANCE MODEL FOR CROP PLANTS

Albert Liptay[1,2] and Jerry Stoller[1]

[1]Stoller Enterprises Inc., 4001 W Sam Houston Pkwy N, Houston, TX 77043 USA

[2]Corresponding author: aliptay@hotmail.com

Stoller had developed and has been using a hormone balance model for the last two decades to understand hormone effects and hormone balance in crop and other plants. This model focuses on the synthesis, transport and use of the hormones in the shoots versus the roots. The model forms the basis of Stoller's knowledge base of crop productivity, product quality and pest tolerance.

CAN THERMODYNAMIC LAWS DRIVE PRODUCT DISCOVERY FOR INCREASED YIELD?

Richard Woodward[1,2] and Jerry Stoller[1]

[1]Stoller Enterprises Inc., 4001 W Sam Houston Pkwy N, Houston, TX 77043 USA

[2]Corresponding author: rwoodward@stollerusa.com

Four laws of thermodynamics define the operation of the natural world. Only recently have we seen in the literature the convergence of these physical laws with specific examples in the biological sciences. Consideration of the four laws of thermodynamics provides a fresh perspective to plant growth and development. However, the botanical literature historically shrouded references to energy transformation and utilization that are useful for thermodynamic interpretations. Discussions of resource allocation, resource partitioning, source-sink relations, remobilization, compensation, adaptation, and hardening all consider energy flux. Energy enters / leaves plant systems as heat, chemical bonds, and electromagnetic radiation. Plant growth regulators, signaling molecules and minerals modulate the passage and utilization of energy plant systems. Evaluation of temporal changes in temperature (delta T) and changes in pressure (delta P) during the growing season reveals that the delta T is largest in the spring, narrows as the season progresses and inverts near harvest. Subtle refinements in delta T that occur in vivo during the season link directly to delta P. Likewise, the magnitude of delta P starts high with winter soil moisture and spring rains, and narrows as the season progresses. We propose a functional relationship of the tenants of thermodynamics for each of the major PGRs and then extend this concept to specific minerals for the major phases of plant growth. This paper assembles clear examples of each thermodynamic law in plant physiology and further proposes by example, a rational approach to the use of these laws for intervention and manipulation of plant growth, development, reproduction, and maturation.

MORPHOLOGICAL AND PHYSIOLOGICAL RESPONSES OF CREEPING BENTGRASS TREATED WITH ROOT MASS 20/20® AND STIMULATE® UNDER HEAT AND DROUGHT STRESS

Marcus Jones[1,3], Jerry Stoller[1], Xunzhong Zhang[2] and Erik Ervin[2]

[1]Stoller Enterprises, Inc., 4001 W Sam Houston Pkwy N, Houston, TX 77043 USA

[2]Department of Crop & Soil Environmental Sciences, Virginia Tech, Blacksburg, VA, USA

[3]Corresponding author: mjones@stollerusa.com

Drought stress confounded by heat stress is a major limiting factor in the culture of creeping bentgrass (*Agrostis stolonifera* L.). The decline in plant vigor is often associated with reduced root growth, photosynthetic rate, and rise in oxidative stress. Biostimulant products have demonstrated the ability to increase turfgrass stress tolerance. Biostimulants often contain phytohormones or substances that alter endogenous hormone levels and stimulate plant growth and metabolism with a response not attributed to mineral nutrition. The objective of this research was to investigate morphological and physiological characteristics of sequential foliar applications of Root Mass 20/20®, a biostimulant applied alone and in combination with Stimulate®, a plant growth regulator (PGR) on creeping bentgrass. Plugs of 'Penn A-4' creeping bentgrass were evaluated in a growth chamber used to impose moderate heat and drought stress. In general, Root Mass 20/20 alone or in combination with Stimulate generally improved root growth, turfgrass quality, photosynthetic rate, and levels of super oxide dismutase. The data indicate excellent potential of Root Mass 20/20 and Stimulate for improving creeping bentgrass heat and drought tolerance.

INTRODUCTION

Creeping bentgrass is a cool-season grass commonly utilized for intensely managed, high value sporting environments. It grows best during the spring and fall months when temperatures are cool and forms a uniform, dense playing surface under favorable environmental conditions. Plant vigor and productivity begin to decline during the summer months, an occurrence referred to as summer bentgrass decline (Dernoeden, 2000). Symptoms of summer bentgrass decline include thinning of the turf canopy, excessive leaf senescence, and root dieback. High temperatures can also reduce tiller density, root growth, and photosynthetic rate of creeping bentgrass (Xu and Huang, 2001). Drought stress confounded by summer heat stress is a major limiting factor in the culture of creeping bentgrass (Beard, 1989).Wang et al. (2003, 2004) demonstrated that creeping bentgrass subjected to high rootzone temperatures experienced shoot injury from a decrease in cytokinin production and initiation of oxidative stress. Similar decreases in turf quality and cytokinin content of shoots and roots have been observed in creeping bentgrass cultured under high soil temperatures and high soil and air temperatures (Liu et al., 2002). High soil temperatures have also been found to hasten leaf senescence in creeping bentgrass, especially when combined with high air temperatures (Liu and Huang, 2002). Exogenously applied cytokinins can help alleviate heat stress injury and improve the quality and growth of creeping bentgrass (Liu and Huang, 2002; Liu et. al., 2002).

Biostimulant products have been shown to increase turfgrass tolerance to various environmental stresses. Biostimulants are products able to stimulate plant growth and metabolism with a response not attributed to mineral nutrition (Schmidt et al., 2003). Seaweed extracts, humic and other organic acids, amino acids, plant hormones, bacteria, fungi, microbes, and organic byproducts are often active ingredients of biostimulants, along with small amounts of mineral nutrition (Karnok, 2000). The objective of this research was to investigate the morphological and physiological effects of Root Mass 20/20 and Stimulate on creeping bentgrass subject to heat and drought stress.

MATERIALS AND METHODS

Established plugs of 'Penn A-4' creeping bentgrass were taken from a mature United States Golf Association (USGA) sand putting green at Virginia Tech Turfgrass Research Center on 11 March 2013 and roots were trimmed at the bottom of the thatch layer at transplant. The plugs were grown in pots (15.2 cm x 12.7cm deep) filled with a USGA medium sand. Pots were arranged in a randomized complete block with four replicates. The grass was grown for 14 days before the initial application of treatments.

Treatments included Root Mass 20/20, a biostimulant and Stimulate, a plant growth regulator (Stoller Enterprises, Inc., Houston, Texas) applied alone and in combination (Table 1). Root Mass 20/20 is a 2N-0P-3K with 5% humic acid originating from leonardite ore. Stimulate contains cytokinin, gibberellic acid, and auxin formulated in a 2:1:1 ratio.

Controlled environment settings were used to impose moderate heat and drought stress (Table 2). A growth chamber malfunction compromised the initial plant samples and a second set of plugs were collected on 28 July 2013. All plants were fertilized with Nutriculture 28-8-18 at 4.9 kg ha^{-1} every 14 days. Additional nitrogen and potassium were applied as needed to balance the nutrients from the 2-0-3 analysis of Root Mass 20/20 and equilibrate the nutrients among all treatments.

Roots were washed to remove excess soil media and dry or ashed root weight was determined. Due to the growth chamber malfunction, root weight was the only data collected during the initial trial. Turf quality was rated based on a 1-9 scale with 9 indicating the best quality. Photosynthetic rate (Pn) was determined using a Licor LI6400XT photosynthesis system. Leaf samples were collected at days 14, 28 and 42 for analysis of chlorophyll content, superoxide dismutase (SOD) activity, abscisic acid (ABA) content, and dehydrin protein expression.

Analysis of variance (ANOVA) was performed using the CAR package and lm, anova, and summary functions within R software (R Development Core Team, 2013) and an alpha of0.10 was used for all mean comparisons.

RESULTS AND DISCUSSION

Root Mass 20/20 applied alone or in conjunction with Stimulate significantly increased 'Penn A-4' creeping bentgrass root weight of spring harvested plugs (Fig. 1A). A 90% increase in root weight resulted from Root Mass 20/20 being applied at 1.17 L ha^{-1} every 7 days compared to control plants. Root Mass 20/20 applied alone at 2.32 L ha^{-1} every 14 days or in combination with Stimulate at 0.16 L ha^{-1} also significantly increased root growth 48 and 55%, respectively.

Changes in root weight during the second trial were more modest. Root Mass 20/20 applied at 2.32 L ha^{-1} every 14 days was the only treatment to significantly increase root weight (Fig. 1B). Xu and Huang (2006) demonstrated that root metabolic activity of creeping bentgrass declined during the summer months resulting in minimal amounts of carbon being allocated to root growth. Because plugs for the second trial were harvested during July, it's possible the bentgrass was in physiological decline and few resources were available for root growth.

In general, 'Penn A-4' creeping bentgrass exhibited an increase in SOD levels at 14 and 28 days into the stress period when treated with Root Mass 20/20 alone or in combination with Stimulate (Fig. 2). Antioxidants, such as SOD, help protect plants from reactive oxygen species which can compromise the cell wall structure of major plant organelles such as chloroplasts and mitochondria which are essential to fundamental plant processes.

A higher Pn of 'Penn A-4' creeping bentgrass was observed throughout the 42 day stress regime when treated with Root Mass 20/20 alone or in combination with Stimulate (Fig. 3). Plants receiving Root Mass 20/20 applied at 1.17 L ha^{-1} every 7 days or a combination of Root Mass and Stimulate, each at 0.16 L ha^{-1} every 14 days exhibited higher Pn rates throughout the trial.

'Penn A-4' creeping bentgrass consistently demonstrated higher turf quality throughout the 42 day stress regime when treated with Root Mass 20/20 alone or in combination with Stimulate (Fig. 4). While the quality of the untreated plants continued to decline in the presence of heat and drought stress, turfgrass quality of the treated plants plateaued at 28 days into the stress regime.

Significant differences were not observed among the treatments for chlorophyll levels and few differences were noted for leaf ABA concentration (Fig. 5 and 6) or dehydrin protein expression (data not shown). A decrease in leaf ABA occurred at 42 days for plants receiving Root Mass 20/20 at the 1.17 L ha^{-1} rate. Wang et al. (2003) and DaCosta and Huang (2007) reported that Kentucky bluegrass (*Poa pratensis* L.) and creeping bentgrass cultivars tolerant of drought exhibited lower ABA accumulation rates than drought-sensitive cultivars during short-term drought stress, suggesting that a low accumulation rate of ABA in leaves would be beneficial for the maintenance of photosynthesis during short-term drought.

In summary, Root Mass 20/20 alone or in combination with Stimulate generally improved 'Penn A-4' turfgrass quality, Pn, and increased levels of SOD. Significant increases in root growth were also observed when treatments occurred prior to the onset of stress. The data suggest that spring applications of Root Mass 20/20 alone or in combination with Stimulate can help the plant establish a large root system. Summer applications during a time of natural metabolic decline can help maintain turfgrass quality through enhanced photosynthesis and SOD production. These data indicate excellent potential of Root Mass 20/20 and Stimulate for improved creeping bentgrass heat and drought tolerance.

LITERATURE CITED

Beard, J.B. 1989. Turfgrass water stress: drought resistance components, physiological mechanisms, and species-genotype diversity. Proc. Int. Turfgrass Res. Conf. 6:23-28.

DaCosta, M. and B. Huang. 2007. Drought survival and recuperative ability of bentgrass species associated with changes in abscisic acid and cytokinin production. J. Amer. Soc. Hort. Sci. 132:60-66.

Dernoeden, P.H. 2000. Creeping bentgrass management: Summer stresses, weeds, and selected maladies. Ann Arbor Press, Chelsea, MI.

Karnok, Keith J. 2000. Promises, promises: can biostimulants deliver? Golf Course Mgt. 68:67- 71.

Liu, X., and B. Huang. 2002. Cytokinin effects on creeping bentgrass responses to heat stress: II. leaf senescence and antioxidant metabolism. Crop Sci. 42(2):466–472.

Liu, X., B. Huang, and G. Banowetz. 2002. Cytokinin effects on creeping bentgrass responses to heat stress: I. shoot and root growth. Crop Sci. 42(2):457–465.

Mato, M.e., L.M. Gonzallez-Alonso and J. Mendy. 1972. Inhibition of enzymatic indoleacetic acid oxidation by fulvic acids. Soil Biology and Biochemistry 4:475-478.

R Core Team. 2013. R: A language and environment for statistical computing. R Foundation for Statistical Computing, Vienna, Austria. URL http://www.R-project.org/.

Schmidt, R.E., E.H. Ervin, and X. Zhang. 2003. Questions and answers about biostimulants. Golf Course Mgt. 71:91-94.

Wang, Z., J. Pote, and B. Huang. 2003. Responses of cytokinins, antioxidant enzymes, and lipid peroxidation in shoots of creeping bentgrass to high root-zone temperatures. J. Am. Soc. Hortic. Sci. 128(5):648–655.

Wang, Z., Q. Xu, and B. Huang. 2004. Endogenous cytokinin levels and growth responses to extended photoperiods for creeping bentgrass under heat stress. Crop Sci. 44(1):209–213.

Xu, Q., and B. Huang. 2001. Morphological and physiological characteristics associated with heat tolerance in creeping bentgrass. Crop Sci. 41(1):127–133.

Xu, Q., and B. Huang. 2006. Seasonal changes in root metabolic activity and nitrogen uptake for two cultivars of creeping bentgrass. Hort. Sci. 41(3):822-826.

Table 1. Treatments evaluated for their ability to alleviate heat and drought stress of 'Penn A-4' creeping bentgrass.

Treatment[z]	Rate ($L \cdot ha^{-1}$)	Application interval (days)
Control	-	-
Mass 20/20[y]	1.17	7
Root Mass 20/20	2.32	14
Root Mass 20/20 + Stimulate[x]	0.16/0.16	14

[z] Additional nitrogen and potassium was applied to equilibrate nutrient among all treatments.

[y] Root Mass 20/20 is a 2-0-3 with 5% humic acid.

[x] Stimulate contains cytokinin, gibberellic acid and auxin.

Table 2. Experimental schedule used to impose heat and drought stress on 'Penn A-4' creeping bentgrass.

Day	-14	1	7	8	14	28	42
Temperature (day/night; °C)	23/16	29/21		35/25[z]			
Soil moisture (pot capacity)	100	50-100		50[y]			
Fertilization[x]	x	x			x	x	x
Treatment	1st app.	Refer to treatment schedule in Table 1					

[z] Gradual temperature increase to 35/25 °C of two degrees per hour.

[y] Pots weighed and watered to 50% pot capacity every other day.

[x] Nitrogen and potassium was applied to equilibrate the nutrients among all treatments.

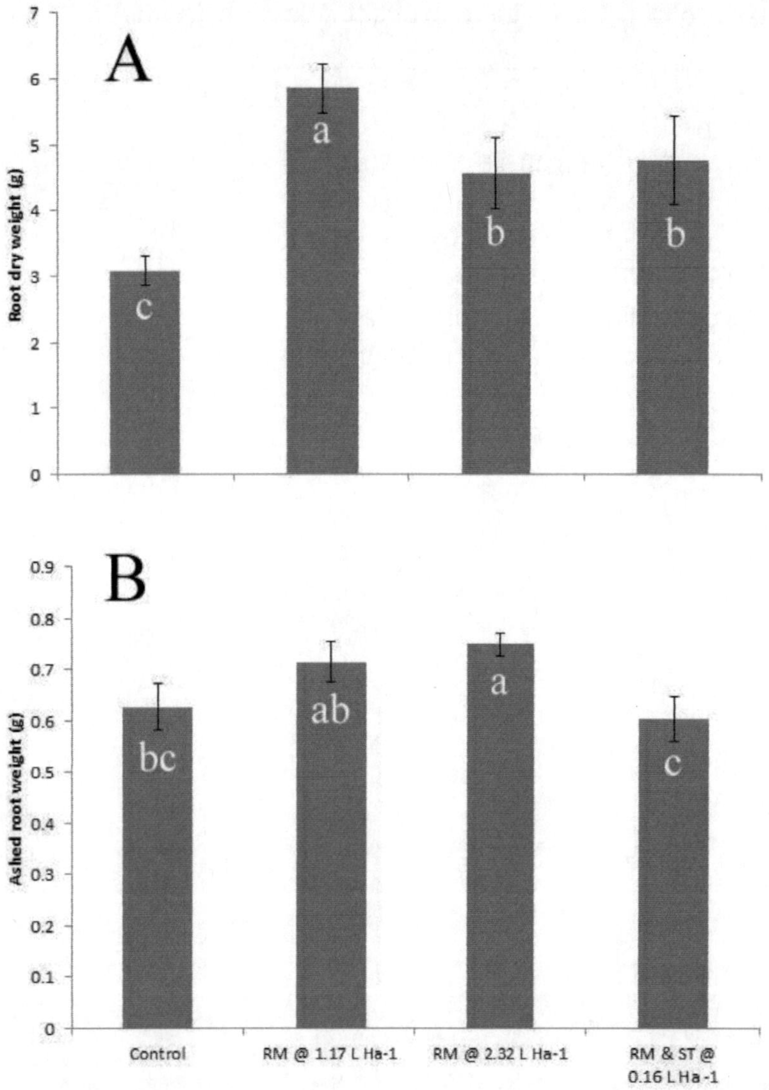

Fig. 1. Root growth of 'Penn A-4' creeping bentgrass collected on 11 March 2013 (A) or 28 July 2013 (B). Plants were subject to heat and drought stress and treated with Root Mass 20/20, Stimulate, or a combination of the two. Error bars represent standard error of the mean (n=4). Bars with the same letters are not significantly different at P = 0.10.

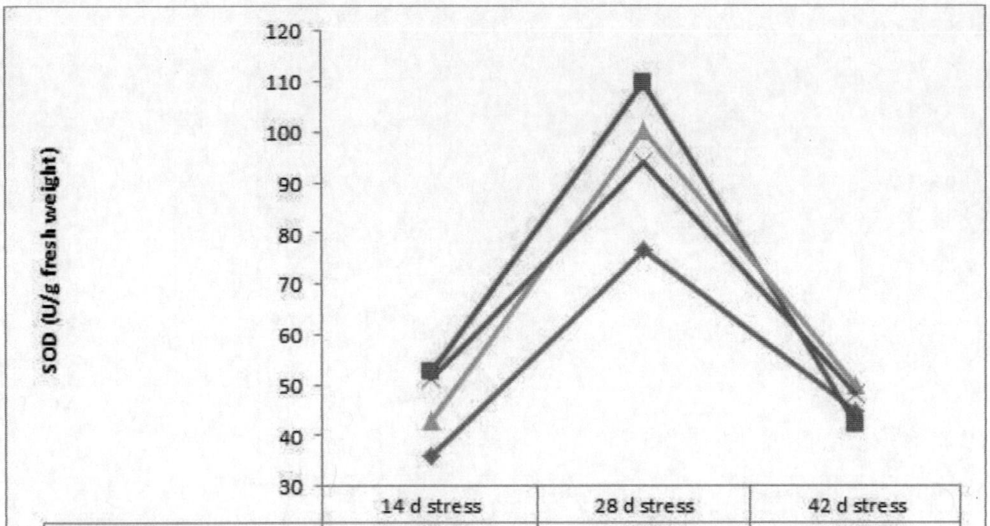

	14 d stress	28 d stress	42 d stress
Control	36.1 b	76.8 b	44.5 ab
RM @ 1.17 L Ha-1	52.6 a	109.6 a	42.1 b
RM @ 2.32 L Ha-1	42.9 ab	100.3 a	49.8 a
RM & ST @ 0.16 L Ha -1	51.4 a	93.9 ab	48.0 ab

Fig. 2. Super oxide dismutase activity of 'Penn A-4' creeping bentgrass subject to heat and drought stress treated with Root Mass 20/20, Stimulate, or a combination of the two. Means followed by same letters within each column are not significantly different at P = 0.10.

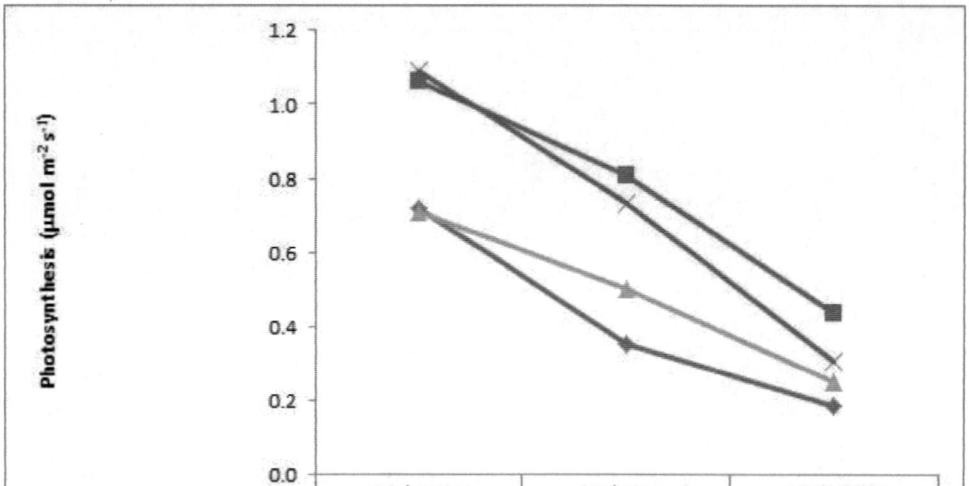

	14 d stress	28 d stress	42 d stress
Control	0.7 b	0.4 b	0.2 c
RM @ 1.17 L Ha-1	1.1 a	0.8 a	0.4 a
RM @ 2.32 L Ha-1	0.7 b	0.5 b	0.2 c
RM & ST @ 0.16 L Ha -1	1.1 a	0.7 a	0.3 b

Fig. 3. Net photosynthetic rate of 'Penn A-4' creeping bentgrass subject to heat and drought stress treated with Root Mass 20/20, Stimulate, or a combination of the two. Means followed by same letters within each column are not significantly different at P = 0.10.

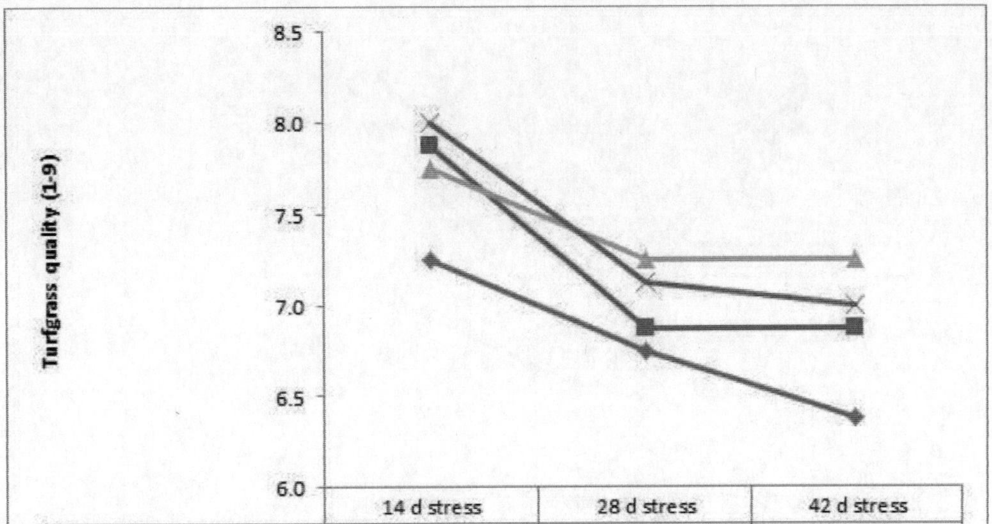

	14 d stress	28 d stress	42 d stress
Control	7.3 b	6.8 b	6.4 c
RM @ 1.17 L Ha-1	7.9 a	6.9 b	6.9 b
RM @ 2.32 L Ha-1	7.8 a	7.3 a	7.3 a
RM & ST @ 0.16 L Ha-1	8.0 a	7.1 a	7.0 ab

Fig. 4. Turfgrass quality of 'Penn A-4' creeping bentgrass subject to heat and drought stress treated with Root Mass 20/20, Stimulate, or a combination of the two. Means followed by same letters within each column are not significantly different at P = 0.10.

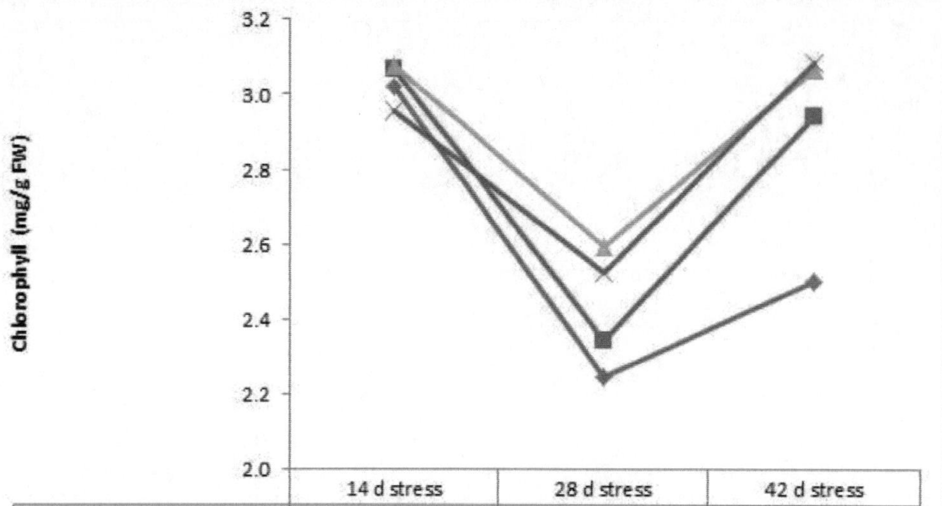

	14 d stress	28 d stress	42 d stress
Control	3.0 a	2.2 a	2.5 b
RM @ 1.17 L Ha-1	3.1 a	2.3 a	2.9 ab
RM @ 2.32 L Ha-1	3.1 a	2.6 a	3.1 ab
RM & ST @ 0.16 L Ha-1	3.0 a	2.5 a	3.1 ab

Fig. 5. Chlorophyll content of 'Penn A-4' creeping bentgrass subject to heat and drought stress treated with Root Mass 20/20, Stimulate, or a combination of the two. Means followed by same letters within each column are not significantly different at P = 0.10.

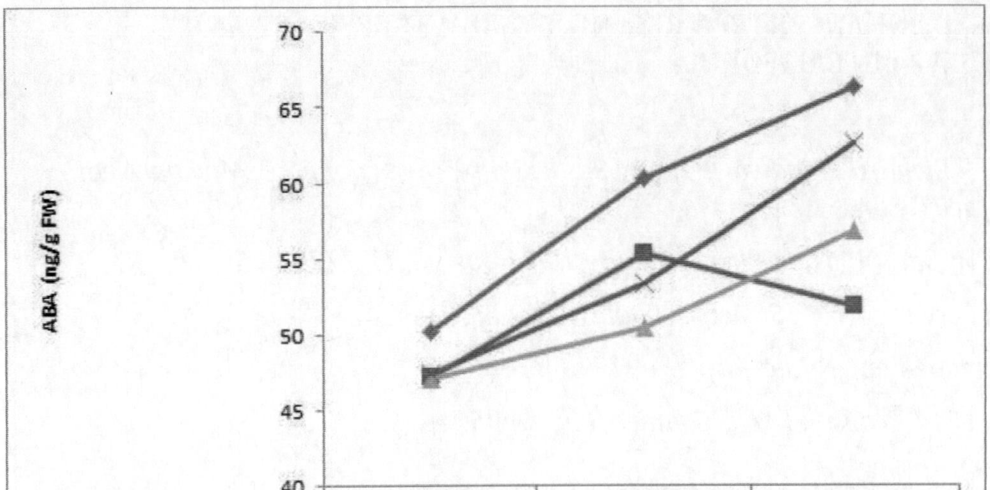

	14 d stress	24 d stress	42 d stress
Control	50.2 a	60.4 a	66.4 a
RM @ 1.17 L Ha-1	47.2 a	55.3 a	51.9 b
RM @ 2.32 L Ha-1	47.1 a	50.4 a	56.9 ab
RM & ST @ 0.16 L Ha-1	47.4 a	53.4 a	62.7 ab

Fig. 6. Leaf abscisic acid of 'Penn A-4' creeping bentgrass subject to heat and drought stress treated with Root Mass 20/20, Stimulate, or a combination of the two. Means followed by same letters within each column are not significantly different at P = 0.10.

STIMULATE™ YIELD ENHANCER (EPA REG. NO. 57538-13) EFFICACY AND PHYTOTOXICITY DATA IN CALIFORNIA

Robert R. Shortell[1,6], Richard Woodward[1], Jerry Stoller[1], Preston Watwood[2], Michael Neal[3], Lance Beem[4] and Jack Devine[5]

[1]Stoller USA, 4001 W. Sam Houston Pkwy N., Suite 100 Houston, TX 77043 USA

[2]General Vineyard Services, 46 4th Street, Gonzales, CA 93926 USA

[3]M&L Vineyard Management, Rutherford, CA 94573 USA

[4]Beem Consulting, 8717 Bronson Drive, Granite Bay, CA 95746 USA

[5]Devine Research

[6]Corresponding author: rshortell@stollerusa.com

Stimulate™ Yield Enhancer is a plant growth regulator and yield enhancer that contains 0.009% cytokinin, as kinetin, 0.005% giberellic acid, as giberellic acid-3, and 0.005% auxin, as indole-3-butyric acid, as the active ingredients. The EPA approved label allows for soil and foliar applications of Stimulate™ Yield Enhancer at two to eight ounce per acre per application on bean, beet, broccoli, Brussels sprout, cabbage, cauliflower, corn, cotton, cucumber, flowering plants, lettuce, melon, onion, orange, ornamentals, peanut, pepper, potato, rice, shrubs, small fruits, sorghum, soybean, squash, strawberry, tobacco, tomato, tree fruits, turfgrass (greens, tees, fairways and sod), vines and wheat. Uses are also approved for seed treatment at two to four ounce per hundred weight and application in transplanting water or through the irrigation system. For best results, Stimulate™ Yield Enhancer should be applied in the early mornings or late evenings.

Reported herein are data from multiple efficacy and phytotoxicity trials conducted in California to demonstrate the performance and safety of Stimulate™ Yield Enhancer on wine grape, nursery stock, pomegranate, olive and turfgrass. Stimulate™ Yield Enhancer should be applied as a foliar spray or through the drip irrigation at two to eight ounce per acre no more than every seven days within the growing season.

During the period 2004-2010, ten efficacy trials were conducted in California to evaluate Stimulate™ Yield Enhancer on wine grape, nursery stock, pomegranate, olive and turfgrass. From one to eight applications (per trial per crop) were evaluated in the ten trials (Table 1). In all ten trials, Stimulate™ Yield Enhancer at two to 16 ounce per acre provided a positive agronomic response over the untreated that ranged from 0.6 to 480%, with all trials contributing to a measurable significant difference across environments. Stimulate™ Yield Enhancer at two ounce per acre provided a positive agronomic response over the untreated that ranged from 2.5 to 70.9% in the two out of ten trials where this rate was evaluated. Stimulate™ Yield Enhancer at four ounce per acre provided a positive agronomic response over the untreated that ranged from0.0 to 200% in the seven out of ten trials where this rate was evaluated. Stimulate™ Yield

Enhancer at six ounce per acre provided a positive agronomic response over the untreated that ranged from 0.0 to 87.5% in the three out of ten trials where this rate was evaluated. Stimulate™ Yield Enhancer at eight ounce per acre provided a positive agronomic response over the untreated that ranged from 0.0 to 371% in the seven out ten trials where this rate was evaluated. Stimulate™ Yield Enhancer at 12 ounce per acre provided a positive agronomic response over the untreated that ranged from 0.0 to 480% in the two out of ten trials where this rate was response over the untreated that ranged from 13.2 to 47.9% in the one trial where this rate was evaluated. At present, the 12 and 16 ounce per acre rate is not included on the Stimulate™ Yield Enhancer label.

Phytotoxicity observations were made in all ten California efficacy trials with Stimulate™ Yield Enhancer on wine grape, nursery stock, pomegranate, olive and turfgrass. As shown in Table 2, no phytotoxicity was observed in any of the trials even where Stimulate™ Yield Enhancer was applied eight times at 16 ounce per acre (double the recommended label rate).

The support data reported herein show that Stimulate™ Yield Enhancer at the label use rates of two to eight ounce per acre per application is very safe to wine grape, nursery stock, pomegranate, olive and turfgrass. The data further show that wine grape, nursery stock, pomegranate, olive and turfgrass treated with Stimulate™ Yield Enhancer show positive plant performance attributes including crop quality and yield response where appropriate.

Table 1. Efficacy performance of Stimulate™ Yield Enhancer from 2004 to 20120 in California.

Location	Crop	No. of applications[z]	Rate[y]	Metrics evaluated[x]
Greenfield	Grape (wine)	8	2, 4, 8, 16	root architecture*, shoot architecture*, vine vigor
Greenfield	Grape (wine)	3	2	berry size, Brix, internode length*, yield
Oakville	Grape (wine)	7	2	bunch shoulder weight, berry size, internode length*, bunch count per vine*, yield*
St. Helena	Grape (wine)	8	2	bunch shoulder weight, berry size, internode length*
Grass Valley	Olive	6	2, 4, 8, 12	rootstock caliper, scion caliper, scion architecture*, root weight*
Grass Valley	Pomegranate	6	2, 4, 8, 12	rootstock caliper, scion caliper, scion architecture*, root weight*
Rio Oso	Turfgrass	1	2, 4, 8	root architecture*, root area*
Rio Oso	Turfgrass	1	2, 4, 8	root architecture*, root area*
Rio Oso	Turfgrass	1	2, 4, 8	root architecture*, root area*
Rio Oso	Turfgrass	1	2, 4, 8	root architecture*, root area*

[z] Total number of applications made.

[y] Values listed as onces per acre.

[x] Asterisk denotes significance at the P = 0.05 level (Student-Newman-Keuls).

Table 2. Phytotoxicity summary of Stimulate™ Yield Enhancer from 2004 to 2010 in California listed as percent increase over the untreated control.

Location	Crop	Trial ID	Rate[z]	No. of applications[y]	Phytotoxicity[x]	Exhibit no.
Greenfield	Grape (wine)	Stim Watwood	16	8	0	1
Greenfield	Grape (wine)	Chardonnay	2	3	0	2
Oakville	Grape (wine)	Merlot	2	7	0	3
St. Helena	Grape (wine)	Semillon	2	8	0	4
Grass Valley	Olive	Nursery Olives	12	6	0	5
Grass Valley	Pomegranate	Nursery Pomegranate	12	6	0	6
Rio Oso	Turfgrass	STOGrnTurf1	8	1	0	7
Rio Oso	Turfgrass	STOGrnTurf2	8	1	0	8
Rio Oso	Turfgrass	STOGrnTurf3	8	1	0	9
Rio Oso	Turfgrass	STOGrnTurf4	8	1	0	10

[z] Values listed as ounces per acre.

[y] Total number of applications made.

[x] Phytotoxicity is visually rated on a 0 to 100% scale where 0 = no plant chlorosis, necrosis or other injury symptoms.

MANIPULATIONS OF CAROTENOID BIOSYNTHESIS IN SPECIALTY CROPS USING UNIQUE PLANT GROWTH REGULATORS

Dean A. Kopsell[1,2], Carl E. Sams[1], and T. Casey Barickman[1]

[1]Plant Sciences Department, The University of Tennessee, 2431 Joe Johnson Drive, Knoxville, TN 37996 USA

[2]Corresponding author: dkopsell@utk.edu

Besides essential vitamins and minerals, specialty vegetable crops provide the human diet with antioxidant phytochemicals. An important class of phytochemicals are the lipid-soluble carotenoid pigments. Health benefits of carotenoids include pro-vitamin A activity, immune system enhancement, and prevention of certain cancers, cardiovascular diseases, and aging eye diseases. Carotenoids are integrated into light harvesting complexes of chloroplasts and function as photo-protectants by quenching free radicals and dissipating excess thermal energy. The carotenoid biosynthetic pathway was elucidated in the mid-1960s. Genes for the major enzymes functioning in carotenoid biosynthesis have been identified and are used to influence biosynthesis through genetic engineering. Manipulation of biochemical fluxes within the pathway can also be accomplished through manipulation of the growing environment and applications of plant growth regulators (PGRs). We have identified keys areas of the biosynthetic pathway impacted through applications of PGRs and other unique biochemical compounds to increase concentrations of important carotenoid pigments, which can be valuable for plant protection or nutritional enhancement of food crops.

INTRODUCTION

Consumption of vegetables provides the human diet with many essential vitamins and minerals important for health maintenance. Vegetables also contain secondary metabolite phytochemicals, which provide benefits beyond normal health maintenance and nutrition and play active roles in chronic disease reductions. An important class of phytochemicals is the carotenoids. Carotenoids are produced in plants via the isoprenoid biochemical pathway. Carotenoids are lipid soluble pigments found in many vegetable crops, which possess reported health benefits of reducing cancers (lycopene), cardiovascular (lycopene), and aging eye diseases (lutein and zeaxanthin) when regularly consumed in the diet. One of the most important physiological functions of carotenoids in human nutrition is as vitamin A precursors (β-carotene). Humans cannot synthesize carotenoids; therefore, fruits and vegetables are primary sources of carotenoids in human diets world-wide (Kopsell and Kopsell, 2010).

Isoprenoid compounds are a diverse group of plant secondary metabolites. Isoprenoids are produced via a common 5-carbon (C_5) isopentenyl diphosphate (IPP) unit and its isomer dimethylallyl diphosphate (DMAPP). Plants synthesize IPP and DMAPP in plastids via the methylerythritol 4-phosphate (MEP) biosyntheitic pathway.

Cytosolic IPP is produced through the mevalonic acid (MVA) pathway (Rodríguez-Concepción and Boronat, 2002). Isoprenoids function in protecting plants against herbivore predation, pathogen attack, as allelopathic chemicals that influence competition from surrounding plants, as well as pollinator attractants. Phytohormones exert numerous important plant physiological responses at different stages of plant development. Several important phytohormones are linked to IPP biosynthesis in plants. In the MVA pathway, farnesyl diphosphate (FPP) is the precursor for sesquiterpens, polyprenols, and phytosteroids. Brassionosteroids are derived from phytosteroids. In the MEP pathway, geranylgeranyl diphosphate (GGPP) is the precursor for gibberellins, carotenoids, phyloquinones, plastoquinones, tocopherols, and chlorophylls. The hormone abscisic acid (ABA) is produced via the carotenoid biosynthetic pathway. Strigolactones are a relatively new class of plant hormones, which inhibit tillering, and shoot branching in plants. Strigolactones are an aopcarotenoid cleavage product of β-carotene in the carotenoid plathway (Alder et al., 2012).

The carotenoid biosynthetic pathway in plants was elucidated in the mid-1960s (Fraser and Bramley, 2004). Carotenoids are produced in the plastids and are derived via the IPP biochemical pathway. In the first step in biosynthesis, isopentenyl diphosphate is isomerized to DMAPP, which becomes the substrate for the C_{20} compound GGPP. The enzyme GGPP synthase catalyses the formation of geranylgeranyl diphosphate from isopentenyl diphosphate and dimethylallyl diphosphate. The first step unique to carotenoid biosynthesis is the condensation of two molecules of GGPP to form the first C40 carotenoid, the colorless phytoene pigment, via phytoene synthase. The carotenoid pathway then branches at the cyclization reactions of lycopene to produce carotenoids with either two β-rings (e.g. β-carotene, zeaxanthin, antheraxanthin, violaxanthin, and neoxanthin) or carotenoids with one β-ring and one ε-ring (e.g. α-carotene and lutein) (Cunningham and Gantt, 1998; Bramley, 2002). The pathway advances with the additions of oxygen moieties, which convert the hydrocarbons, α-carotene and β-carotene, into the oxygenated subgroup referred to as the xanthophylls. Further steps in xanthophyll synthesis include epoxydation reactions. The reversible epoxidation/de-epoxidation reaction converting violaxanthin back to zeaxanthin via the intermediate antherxanthin is collectively referred to as the violaxanthin cycle and is vital for energy dissipation from incoming solar radiation (Fraser and Bramley, 2004; Bramley, 2002). One important regulator that coordinates response to environmental stress is the hormone ABA, which is synthesized from xanthophyll pigments neoxanthin and violaxanthin at the end of the carotenoid biosynthetic pathway (Taylor et al., 1988).

The role of ABA in protecting the xanthophyll cycle [de-epoxidation of violaxanthin (VIO) to zeaxanthin (ZEA)] and the photosynthetic apparatus from photooxidative stress is well documented (Ederli, et al., 1997; Du et al., 2010; Galvez-Valdivieso et al., 2009). Exogenous applications of ABA to barley (*Hordeum vulgare*) seedlings resulted in partial protection of photosystem II (PSII) against photoinhibition at low temperatures; moreover, total carotenoid and xanthophyll carotenoid concentrations in the seedlings increased by 122% (Ivanov et al., 1995). Haisel et al. (2006) found that seedlings of bean (*Phaseolus vulgaris*), tobacco (*Nicotiana tabacum*), beets (*Beta vulgaris*), and corn (*Zea mays*) pre-treated with ABA demonstrated increased chlorophyll and carotenoid concentrations under water stress. There is clear evidence of the connection between carotenoid biosynthesis and the metabolism of many plant phytohormones. Work by our group set out to determine the potential to manipulate carotenoid biosynthesis through exogenous applications of unique plant growth regulators.

Application of Apo-carotenoids as Unique pgrs. Cleavage of specific carotenoids will produce unique apo-carotenoids that are vital in the biological functions of plants and animals. The most widely recognized apo-carotenoids are retinal (vitamin A) and ABA. The ionones (α-ionone and β-ionone) are cleavage products of carotene carotenoids that play important roles in the flavor and aroma of fruits, vegetables and ornamental plants (Ohmiya, 2009) (Figure 1). Both α-ionone and β-ionone are also used in the food and cosmetic industries to provide unique sensory qualities to commercial products (Plotto et al., 2006). The health attributes of carotenoids are well established; moreover, epidemiological evidence is showing the ionones to possess unique antioxidant and anticancer properties (Beekwilder et al., 2008).

Carotenoid cleavage dioxygenase (CCD) enzymes regulate carotenoid synthesis and degradation in photosynthetic and non-photosynthetic tissues. Carotenoid accumulation and pools of carotenoids present are determined, in part, by the rate of degradation by CCD enzymes. Deposition, storage, and sequestration of carotenoids will vary depending on the type of plastid. Therefore chloroplasts and chromoplasts will differ in the ability to accumulate carotenoids (Cazzonelli and Pogson, 2010).

Our research group set out to determine the potential to regulate carotenoid concentrations through applications of commercially available α-ionone and β-ionone to potentially increase the nutritional value of specialty vegetable crops. A series of experiments were designed to measure impacts of exogenous applications of ionone compounds on the accumulation and distribution of carotene and xanthophyll carotenoids among leaf, fruit, and root-type tissues. Foliar and soil drench applications of α-ionone and β-ionone had little impact of the accumulation of carotenes within the leaf tissues of basil (*Ocimum basilicum* L.) and kale (*Brassica oleracea* L. var. *acephala* DC); however, foliar applications of β-ionone significantly increased tomato (*Solanum lycopersicum* L.) fruit tissue β-carotene. Soil drenches of α-ionone and β-ionone also increased β-carotene in carrot (*Daucus carota* L. var. *sativa*) root tissues.

Results from our studies indicate the potential to enhance, or possibly retain, tissue β- carotene concentrations through simple applications of apo-carotenoids. Results also show that ionone applications have a higher impact on chromoplast carotenoids. Application of ionone compounds may be a viable method to improve β-carotene concentrations in fruiting and root-type specialty vegetable crops.

Several studies demonstrate exogenously applied ABA can increase chlorophylls and carotenoids in light stressed plants. We sought to determine dose-response effects of ABA in solution culture for maximum carotenoid accumulations in leaf and fruit tissues in two distinct genotypes of dwarf tomato. Tomatoes were grown in controlled environments using solution culture techniques. ABA treatments (s-ABA; Valent BioSciences, Libertyville, IL) were applied to nutrient solutions at concentrations of 0.0, 0.5, 5.0, and 10.0 mg per L. The ABA treatments increased the accumulation of β-carotene, zeaxanthin, lutein, and neoxanthin carotenoids in 'MicroTina' tomato leaf tissue when compared to the control treatment (Barickman et al., 2014) (Table 1).

In 'MicroTina' tomato fruit, there was an increase in lycopene concentrations. Lycopene increased by 35.5% when comparing ABA treatments concentrations to the control treatment with 0.0 mg per L of ABA. In contrast, there were no significant differences in 'MicroGold' fruit tissue carotenoids among the ABA treatments (Barickman et al., 2014). The impact of ABA on tomato leaf tissue carotenoids may be a result of the indirect regulation of carotenoid biosynthesis by increasing the activity of key enzymes, such as β-carotene hydroxylase and phytoene synthase (Meier et al., 2011).

Application of GA-inhibiting compounds as Unique pgrs. Gibberellins share a common biosynthetic precursor with carotenoid pigments in GGPP. The commercial product Sumagic® (Valent BioSciences, Libertyville, IL) acts to inhibit gibberellin biosynthesis in plants. The product is labeled for use on solanaceous vegetable crop transplants to shorten plants and thicken stems prior to field planting. The active ingredient in Sumagic® is Uniconazole-P. It inhibits endogenous biosynthesis of gibberellic acid by inhibiting P450 *ent*-kaurene oxidase, which catalyzes the oxidation of *ent*-kaurene to *ent*-kaurenoic acid. It can also reduce ABA catabolism to cause accumulation of ABA (Saito et al., 2006). We were interested to see if the gibberellin biosynthesis inhibitor would impact carotenoid physiology by causing a feedback mechanism to move metabolic energy to carotenoid biosynthesis.

Chinese kale (*Brassica oleracea* var. *alboglabra*) was grown in a controlled environment for 30 days. A single foliar application of uniconazole-P was made to the plants at 0, 2, 4, 8, and16 mg per L. The plants were harvested 14 days after treatment. Plant heights at harvest were 13.48, 7.56, 6.94, 6.77, and 5.89 cm for the uniconazole-P treatments of 0, 2, 4, 8, and 16 mg per L, respectively. Plant fresh weight for the shoot tissues at harvest were 16.62, 10.42, 9.72, 8.27, and 8.00 g per plant for the uniconazol-P treatments of 0, 2, 4, 8, and 16 mg per L, respectively. Uniconazole-P reduced plant height and shoot tissue fresh weight as expected. However, shoot tissue carotenoid pigments decreased with increasing concentrations of uniconazole-P. Shoot tissue lutein concentrations were 3.67, 2.68, 2.34, 2.11, and 2.08 mg per 100 g fresh weight for the uniconazole-P treatments of 0, 2, 4, 8, and 16 mg per L, respectively. Shoot tissue β-carotene concentrations were 1.67, 1.24, 1.09, 0.93, and 0.92 mg per 100 g fresh weight for the uniconazole-P treatments of 0, 2, 4, 8, and 16 mg per L, respectively. Uniconazole-P blocked GA metabolism and reduced plant height and biomass, but it did not result in redistribution of metabolites within IP pathway.

Conclusions

We have demonstrated that several unique growth regulators impact the carotenoid pathway and thus influence the concentration of carotenoids important in human nutrition and health. Manipulation of fluxes of metabolites within the carotenoid biosynthetic pathway can be accomplished through applications of apo-carotenoid compounds and unique PGRs. We have identified keys areas of the biosynthetic pathway having the potential for manipulation to increase concentrations of important carotenoid pigments. Increasing carotenoid concentrations in specialty vegetable crops may be valuable for nutritional enhancement of food crops.

LITERATURE CITED

Alder, A., M. Jamil, M. Marzorati, M. Bruno, M. Vermathen, P. Bigler, S. Ghisla, H. Bouwmeester, P. Beyer, and S. Al-Babili. 2012. The path from β-carotene to carlactone, a strigolactone-like plant hormone. Science 335:1348-1351.

Barickman, T.C., D.A. Kopsell, and C.E. Sams. 2014. Abscisic acid increases carotenoid and chlorophyll in leaves and fruit of two tomato genotypes. J. Amer. Soc. Hort. Sci. 139(3):261-266.

Beekwilder, J., I.M. van der Meer, A. Simic, J. Uitdewilligen, J. van Arkel, R.C.H. de Vos, H. Jonker, F.W.A. Verstappen, H.J. Bouwmeester, O. Sibbesen, I. Qvist, J.D. Mikkelsen, and R.D. Hall. 2008. Metabolism of carotenoids and apocarotenoids during ripening of raspberry fruit. BioFactors 34:57-66.

Bramley, P.M. 2002. Regulation of carotenoid formation during tomato fruit ripening and development. J. Expt. Bot. 53(377):2107-2113.

Cazzonelli, C.I., and B.J. Pogson. 2010. Source to sink: regulation of carotenoid biosynthesis in plants. Trends Plant Sci. 15:266-274.

Cunningham, F.X., and Gantt, E. 1998. Genes and enzymes of carotenoid biosynthesis in plants. Ann. Rev. Plant Physiol. Plant Mol. Biol. 49:577-583.

Du, H., N.L. Wang, F. Cui, X.H. Li, J.H. Xiao, and L.Z. Xiong. 2010. Characterization of the beta-carotene hydroxylase gene *dsm2* conferring drought and oxidative stress resistance by increasing xanthophylls and abscisic acid synthesis in rice. Plant Physiol. 154:1304-1318.

Ederli, L., S. Pasqualini, P. Batini, and M. Antonielli. 1997. Photoinhibition and oxidative stress: effects on xanthophyll cycle, scavenger enzymes and abscisic acid content in tobacco plants.

Plant Physiol. 151:422-428.Fraser, P.D., and Bramley, P.M. 2004. The biosynthesis and nutritional uses of carotenoids. Progress Lipid Res. 43:228-265.

Galvez-Valdivieso, G., M.J. Fryer, T. Lawson, K. Slattery, W. Truman, N. Smirnoff, T. Asami, W.J. Davies, A.M. Jones, N.R. Baker, and P.M. Mullineaux. 2009. The high light response in arabidopsis involves aba signaling between vascular and bundle sheath cells. Plant Cell 21:2143-2162.

Haisel, D., J. Pospisilova, H. Synkova, R. Schnablova, and P. Batkova. 2006. Effects of abscisic acid or benzyladenine on pigment contents, chlorophyll fluorescence, and chloroplast ultrastructure during water stress and after rehydration. Photosynthetica 44:606-614.

Ivanov, A.G., M. Krol, D. Maxwell, and N.P.A. Huner. 1995. Abscisic-acid induced protection against photoinhibition of PSII correlates with enhanced activity of the xanthophyll cycle. FEBS Letters 371:61-64.

Kopsell, D.A. and D.E. Kopsell. 2010. Carotenoids in vegetables: biosynthesis, occurrence, impacts on human health, and potential for manipulation, pp. 645-662. In: R. Ross and V.R. Preedy (eds.). Bioactive Food in Promoting Health: Fruits and Vegetables. Elsevier, St. Louis, Mo. 737 pp.

Meier, S., O. Tzfadia, R. Vallabhaneni, C. Gehring, and E.T. Wurtzel. 2011. A transcriptional analysis of cartoenoid, chlorophyll and plastidial isoprenoid biosynthesis genes during development and osmotic stress repsonses in *Arabidopsis thaliana*. BMC Syst. Bio. 5:77-96.

Ohmiya, A. 2009. Carotenoid cleavage dioxygenases and their apocarotenoid products in plants. Plant Biotech. 26:351-358.

Plotto, A. K.W. Barnes, and K.L. Goodner. 2006. Specific anosmia observed for beta-ionone, but not for alpha-ionone: Significance for flavor research. J. Food Sci. 71:S401-S406.

Rodríguez-Concepción, M. and A. Boronat. 2002. Elucidation of the methylerythritol phosphate pathway for isoprenoid biosynthesis in bacteria and plastids. A metabolic milestone achieved through genomics. Plant Physiol. 130:1079-1089.

Saito, S., M. Okamoto, S. Shinoda, T. Kushiro, T. Koshiba, Y. Kamiya, N. Hirai, Y. Todoroki, Sakata, E. Nambara, and M. Mizutani. 2006. A plant growth retardant, uniconazole, is a potent inhibitor of ABA catabolism in Arabidopsis. Biosci. Biotechnol. Biochem. 70:1731- 1739.

Sharma, P.K., S. Sankhalkar, and Y. Fernandes. 2002. Possible function of ABA in protection against photodamage by stimulating xanthophyll cycle in sorghum seedlings. Current Sci. 82:167-171.

Taylor, I.B., R.S.T. Linforth, R.J. Alnaieb, W.R. Bowman, and B.A. Marples. 1988. The wilty tomato mutants *flacca* and *sitiens* are impaired in the oxidation of ABA-aldehyde to ABA. Plant Cell Environ. 11:739-745.

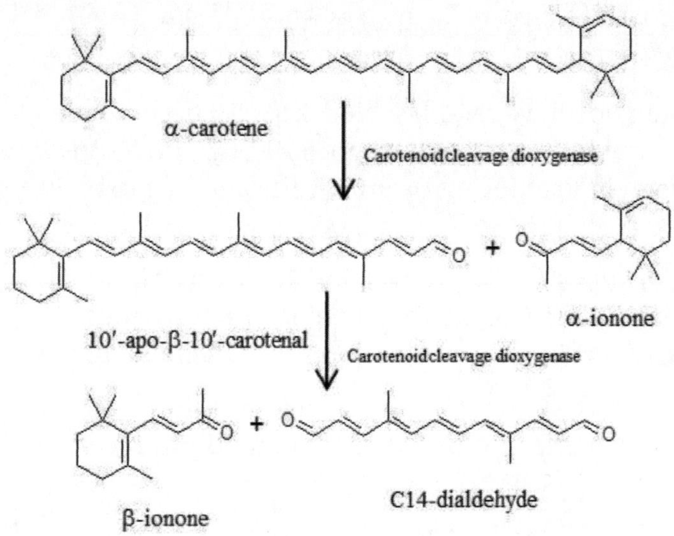

Figure 1. Cleavage of α-carotene to produce α-ionone and β-ionone.

Table 1. Mean values for carotenoid leaf tissue pigment (mg per 100 g fresh weight) for 'MicroTina' tomato plants grown in hydroponic nutrient solution under increasing ABA concentrations.

	Mean carotenoid concentration (mg/100 g fresh weight)[y]						
ABA ($mg \cdot L^{-1}$)[z]	BC	LUT	ZEA	ANTH	NEO	VIO	Total CAR
0.0	3.33	10.95	0.20	1.61	4.28	1.36	20.52
0.5	6.18	15.43	0.35	1.72	6.17	1.82	30.15
5.0	6.79	16.68	0.63	1.90	6.36	1.31	31.82
10.0	6.65	16.46	0.72	1.91	6.20	1.62	31.29
Contrast Control vs. ABA[x]	*	**	**	ns	*	ns	*

[z] Mean separation of the ABA treatments were not significant therefore ABA treatments were pooled for statistical analysis.

[y] BC = β-carotene; LUT = Lutein; ZEA = Zeaxanthin; ANTH = Antheraxanthin; NEO = Neoxanthin; VIO = Violaxanthin; Total CAR = Total carotenoids.

[x] ns, *, **, and *** indicate nonsignificant or significant at $P \leq 0.05, 0.01, 0.001$, respectively.

Effects of Abscisic Acid on Tomato Fruit Aroma Volatiles

T. Casey Barickman[1,2,3], Dean A. Kopsell[1], and Carl E. Sams[1]

[1]Plant Sciences Department, The University of Tennessee, 2431 Joe Johnson Drive, Knoxville, TN 37996 USA

[2]Corresponding author: cb2407@msstate.edu

[3]Current Address: Mississippi State University, North Mississippi Research and Extension Center, Verona, MS 38879.

Aroma volatiles are derived from a diverse set of precursors, such as amino acids, fatty acids and carotenoids in tomato fruit. Many of these volatiles enhance the main flavor components in the fruit, particularly soluble sugars and organic acids. ABA is derived from the carotenoid pathway and there may be an indirect connection to flavor volatiles through this pathway. Therefore, the purpose of this study was to examine the influence of ABA on tomato fruit aroma volatiles. This study identified five flavor volatile compounds that were consistently present in 'Mt. Fresh Plus' tomato fruit tissue. They were 2-methyl furan, (E)-2-hexeanl, 1- hexanol, hexenal, and 6-methyl-5-hepten-2-one. ABA treatments did not have an effect on aroma volatile concentrations in 'Mt. Fresh Plus' tomato fruit. Majority of the volatiles identified did not differ between the ABA treated plants and the ABA control plants. However, ABA treatments did significantly decrease (E)-2-hexenal.

INTRODUCTION

For several decades tomato cultivar selections from plant breeders have emphasized grower demands for yield, fruit size, firmness and resistance to biotic and abiotic diseases (Maul et al., 2000). For the most part, growers get paid for pounds of product in the box and not for taste quality. As a result, the sensory aspects of fruit quality, such as flavor and aroma, have diminished. Consumers frequently associate newer hybrid tomato cultivars with poor flavor and aroma (Klee and Tieman, 2013). In recent years there has been a reconnection between the consumer, producers and breeders to bring to the forefront the flavor and aroma of the tomato fruit. Flavor is a function of aroma volatiles that enhance the flavor quality and removing them greatly reduces flavor intensity (Baldwin et al., 2008).

The chemicals that contribute to the flavor of tomato fruit have been well documented. Approximately 400 volatile compounds have been identified in tomato fruit (Petro-Turza, 1986, Baldwin et al., 2000). Aroma volatiles are derived from a diverse set of precursors, such as amino acids, fatty acids and carotenoids (Klee and Tieman, 2013). The main function of aroma volatiles in tomato fruit is to enhance the main flavor components, which are soluble sugars and organic acids (Klee and Tieman, 2013, Tieman et al., 2012). Therefore, aroma volatiles will enhance sweetness, acidity, or green flavor depending on the perceptions of consumer preferences.

There is a high variation in the profile of aroma volatiles across tomato cultivars (Tieman et al., 2012). Despite the fact that there are 400 volatile compounds known for tomato fruit, some are more abundant than others. Previous research has demonstrated that the more prevalent aroma volatiles in tomato fruit are hexenal, (E)-2-hexenal and 6-methyl-5-hepten-2- one (Buttery and Ling, 1993, Baldwin et al., 1991b). In addition, research has also shown that volatiles cis-3-hexenal, trans-2-hexenal, hexanal and 2-isobutylthiazole contribute to the quality of ripe tomato fruit (Stone et al., 1975) by enhancing the fresh flavor and aroma. Carbonell- Barrachina et al. (2006) demonstrated a difference between different types of tomatoes based on their flavor volatile composition. The tomato cultivar "De la Pera", which contained the highest content of flavor volatiles, received the highest values of odor and aroma. Additionally, not all aroma volatile compounds contribute to tomato flavor equally. For example, a more common aroma volatile, a six carbon hexanal, contribution to tomato flavor more than some other aroma volatile, such as geranial (Klee and Tieman, 2013). The profile of aroma volatiles in any particular tomato fruit will depend on numerous environmental factors, such as temperature, light, seasonal differences and site variations (Tieman et al., 2012).

There are no published reports on the effect of ABA on aroma volatiles. However, ABA is derived from the carotenoid pathway and there may be an indirect connection to flavor volatiles through this pathway. Therefore, the purpose of this study was to examine the influence of ABA on tomato fruit quality, specifically aroma volatiles. Since aroma volatiles are derived from a diverse set of precursors, including amino acids, fatty acids and carotenoids, the assumption is that ABA may positively affect them as well. In addition, aroma volatiles in newer tomato cultivars have not been studied extensively. Therefore, this study further examines this tomato cultivar's fruit quality by identifying its aroma volatiles.

METHODS AND MATERIALS

Plant Culture and Harvest. Seeds of 'Mountain Fresh Plus' tomato (Johnny's Selected Seed, Waterville, ME) were sown into Pro-Mix BX soilless medium (Premier Tech Horticulture, Québec, Canada) and germinated in the greenhouse conditions (Knoxville, TN; 35°N Lat.) at 25/20 °C (day/night) under a 16 h supplemental light at an average of 925 $\mu mol \cdot m^{-2} \cdot s^{-1}$. At 30 days after seeding, the plantlets were transferred to 11-L Dutch pots (Tek Supply, Dyersville, IA) filled with Sunshine® Pro Soil Conditioner (Sungro Horticulture, Agawam, MA). Tomato plants were grown hydroponically with a tomato fertilization program developed at the University of Tennessee (Knoxville, TN). Elemental concentrations of the nutrient solutions were ($mg \cdot L^{-1}$): nitrogen (N; 180), phosphorus (P; 93.0), potassium (K; 203.3), magnesium (Mg; 48.6), sulfur (S; 96.3), iron (Fe; 1.0), boron (B; 0.25), manganese (Mn; 0.25), zinc (Zn; 0.025), copper (Cu; 0.01), and molybdenum (Mo; 0.005). Experimental design was a randomized complete block with a 3 x 4 factorial arrangement of treatments which consisted of six blocks and two replications of each treatment, with individual pots representing an experimental unit. Calcium was applied via the irrigation lines at 60, 90, or 180 mg $Ca \cdot L^{-1}$. ABA treatments were applied as a combination of foliar sprays and root applications. For foliar ABA applications, treatments consisted of DI water control (0.0 mg $ABA \cdot L^{-1}$) or 500 mg $ABA \cdot L^{-1}$. For ABA root applications, treatments consisted of a DI water control (0.0 mg ABA·L^{-1}) or 50 mg ABA·L^{-1} applied via the irrigation lines. ABA spray treatments were applied once weekly till dripping from the foliage, while root applications were applied four times per day with the irrigation cycle. Fruit tissues were harvested 84-90 days after seeding. Fruit from each treatment were separated by replication and were weighed for biomass. At least three fruit from the second clusters for each experimental unit were subsampled, combined, and frozen in liquid nitrogen. Harvested fruit samples were stored at -80 °C prior to analysis.

Tomato Aroma Volatiles: Tomato fruit aroma volatiles were extracted from fresh-frozen red ripe fruit tissues and quantified according to the methods of Baldwin et al. (1991a) with slight modifications. Fruit was removed from -80 °C and thawed until slightly friable. A sample of red ripe fruit from each treatment/replication was blended to a slurry. A 5 mL subsample of the slurry was placed into a 20 mL headspace vial and 2 mL of a saturated $CaCl_2$ solution was added. Sample vials were vortexed for 1 min then allowed to set for 30 min. Samples were placed on a static headspace analyzer (Agilent Technologies, Palo Alto, CA) prior to gas chromatography mass spectroscopy (GC-MS) analysis.

An Agilent 6890 series GC unit with a 5873 mass spectrometer detector (Agilent Technologies) was used to identify tomato aroma volatiles following the method of Baldwin et al. (1991). Chromatographic separations were achieved using an analytical scale (0.25 mm i.d. x 30 m) 0.25 μm DB-wax column (JW; Agilent Technologies), which allowed for effective separation of chemically similar volatile compounds. The peak assignment for individual volatiles was performed by comparing retention times and mass spectra (NIST, 2002) using external standards (hexenal, (E)-2-hexenal, 1-hexanol, 2-methyl furan, and 6-methyl-5-hepten-2- one, (Z)-3-hexenal, isobutyl-2-heptenone, dimethylulfide, 1-penten-3-one; (Acros Organics, Fisher Scientific, Pittsburg, PA).

Results

This study found five aroma volatile compounds that were consistently identified in "Mt. Fresh Plus" tomato fruit tissue produced in the ABA and calcium treatments. These aroma volatiles were 2-methyl furan, (E)-2-hexeanl, 1-hexanol, hexenal, and 6-methyl-5-hepten-2-one (Table 1; Table 2). Several flavor volatiles were identified in tomato fruit but were not present at detectable levels consistently enough to analyze statistically. These compounds were acetone, (Z)-3-hexenal, isobutyl-2-heptenone, dimethylulfide, 1-penten-3-one, dimethyl disulfide and 2- nonynoic acid (Data not shown).

The statistical analysis of the results indicated that there was no interaction between ABA and Ca treatments on tomato fruit tissue. Therefore, the following results are presented separately for ABA and Ca effects. The application of ABA either as a foliar spray (500 mg·L^{-1}), root application (50 mg·L^{-1}), or a foliar spray and root combination application did not significantly affect tomato aroma volatile concentrations of most compounds in tomato fruit tissue overall (Table 1). However, the application of ABA did significantly decrease (E)-2- hexenal (Table 1). ABA applications decreased (E)-2-hexenal by 53.8% when comparing the ABA control treatment to the combination of foliar spray and root treatment combination. Ca treatments did not have a significant effect on tomato fruit volatile concentrations (Table 2).

Discussion

This study examined the effects of ABA and Ca treatments on tomato fruit aroma volatiles. Five aroma volatiles that were identified in tomato fruit correspond to results reported in other studies (Buttery and Ling, 1993, Baldwin et al., 2000, Baldwin et al., 2008). Previous published research has not profiled the fruit aroma volatiles of 'Mt. Fresh Plus' tomato cultivar. However, Baldwin et al. (1991b) studied 6 tomato cultivars and identified a similar profile of aroma volatiles as the current study. One of the most prevalent groups of volatiles identified was aldehyde. Hexenal, one of the major aldehydes in tomato fruit, is considered to be important for fresh tomato flavor (Petro-Turza, 1986) and it significantly contributes to tomato fruit flavor (Buttery and Ling, 1993). Therefore, despite differences in tomato cultivars there are several aroma volatiles that are key components contributing to distinct tomato flavor.

This study found that ABA treatments did not affect most of the aroma volatile concentrations in 'Mt. Fresh Plus' tomato fruit. The majority of the volatile concentrations did not differ from the ABA control treatment. However, ABA treatments did significantly decrease (E)-2-hexenal. These results indicated that ABA treatments may not be conducive for positively affecting the aroma volatile profile of the fruit. In fact, these results indicate that ABA treatments could negatively affect certain aroma volatiles, but do not have a significant overall impact on most aroma volatiles. Different tomato cultivars have varying aroma volatile profiles. Therefore, future research on ABA treatments should compare the 'Mt. Fresh Plus' tomato cultivar to other tomato cultivars to gain a better understanding of ABA's effects on tomato aroma volatiles.

Additionally, results of this study indicated that decreasing Ca concentration did not affect tomato aroma volatiles. This may indicate that tomato aroma is not affected in Ca deficient environments. Therefore, despite the fact that Ca deficiencies did not have an effect on tomato fruit aroma volatiles and carotenoids, overall tomato flavor is still negatively affected since soluble sugars and organic acid are the main components that add flavor to the fruit.

Literature Cited

Baldwin, E. A., Goodner, K. and Plotto, A. 2008. Interaction of volatiles, sugars, and acids on perception of tomato aroma and flavor descriptors. J. Food Sci. 73:S294-S307.

Baldwin, E. A., Nisperos-Carriedo, M. O. and Moshonas, M. G. 1991a. Quantitative analysis of flavor and other volatiles and for certain constituents of two tomato cultivars during ripening. J. Am. Soc. Hort. Sci.116:265-269.

Baldwin, E. A., Nisperoscarriedo, M. O., Baker, R. and Scott, J. W. 1991b. Quantitative analysis of flavor parameter in 6 Florida tomato cultivars (*Lycopersicon esculentum* Mill.). J. Agr. Food Chem 39:1135-1140.

Baldwin, E. A., Scott, J. W., Shewmaker, C. K. and Schuch, W. 2000. Flavor trivia and tomato aroma: Biochemistry and possible mechanisms for control of important aroma components. HortScience 35:1013-1022.

Buttery, R. G. and Ling, L. C. 1993. Volatile Components of Tomato Fruit and Plant Parts. *Bioactive Volatile Compounds from Plants.* American Chem Soci.

Carbonell-Barrachina, A. A., Agusti, A. and Ruiz, J. J. 2006. Analysis of flavor volatile compounds by dynamic headspace in traditional and hybrid cultivars of Spanish tomatoes. Euro. Food Res. Technol. 222:536-542.

Klee, H. J. and Tieman, D. M. 2013. Genetic challenges of flavor improvement in tomato. Trends Gen. 29:257-262.

Maul, F., Sargent, S. A., Sims, C. A., Baldwin, E. A., Balaban, M. O. and Huber, D. J. 2000. Tomato Flavor and Aroma Quality as Affected by Storage Temperature. J. Food Sci. 65:1228-1237.

Petro-Turza, M. 1986. Flavor of tomato and tomato products. Food Rev. Int. 2:309-351.

Stone, E. J., Hall, R. M. & Kazeniac, S. J. 1975. Formation of aldehydes and alcohols in tomato fruit from U-C-14-labeled linolenic and linoleic acids. J. Food Sci. 40:1138-1141.

Tieman, D., Bliss, P., Mcintyre, L. M., Blandon-Ubeda, A., Bies, D., Odabasi, A. Z., Rodriguez, G. R., Van Der Knaap, E., Taylor, M. G., Goulet, C., Mageroy, M. H., Snyder, D. J., Colquhoun, T., Moskowitz, H., Clark, D. G., Sims, C., Bartoshuk, L. and Klee, H. J. 2012. The Chemical Interactions Underlying Tomato Flavor Preferences. Cur. Biol. 22:1035-1039.

Table 1. Mean values of tomato fruit aroma volatiles in fruit of 'Mt. Fresh Plus' tomato plants grown in a greenhouse and treated with Ca in the hydroponic fertilizer solution.

	Conc. (μmol·g⁻¹) fresh weight[z]				
ABA[y]	2-Methyl Furan	(E)-2-Hexenal	6-Methyl-5-Hepten-2-one	1-Hexanol	Hexenal
Control	1.17	462.28	212.86	2.31	133.37
Spray	0.99	381.64	145.54	1.52	125.47
Root	0.97	259.12	155.03	1.40	109.07
Spray/Root	1.03	213.49	210.04	1.13	93.31
P-value[x]	ns	**	ns	ns	ns

[z] The SE of the mean for 2-Methyl Furan ± 0.36; (E)-2-Hexenal ± 81.93; 6-Methyl-5-Hepten-2-one ± 54.32; 1-Hexanol ± 0.81; Hexanol ± 19.46.

[y] ABA treatments control (0.0 mg·L^{-1}); spray (500 mg·L^{-1}); root (50 mg·L^{-1}); spray/root (500 mg·L^{-1}/50 mg·L^{-1}).

[x] ns, *, ** and *** indicate non-significant or significant at P ≤ 0.05, 0.01, 0.001, respectively.

Table 2. Mean values for tomato fruit aroma volatiles of 'Mt. Fresh Plus' tomato plants grown in a greenhouse and treated with Ca in the hydroponics fertilizer solution.

	Conc. ($\mu mol \cdot g^{-1}$) fresh weight[z]				
Ca ($mg \cdot L^{-1}$)	2-Methyl Furan	(E)-2-Hexenal	6-Methyl-5-Hepten-2-one	1-Hexanol	Hexenal
60	0.95	295.63	156.46	1.21	112.15
90	1.41	335.43	190.00	2.41	127.98
180	0.76	356.34	196.14	1.13	105.79
P-value[y]	ns	ns	ns	ns	ns

[z] The SE of the mean for 2-Methyl Furan ± 0.34; (E)-2-Hexenal ± 77.60; 6-Methyl-5-Hepten-2-one ± 49.98; 1-Hexanol ± 0.69; Hexanol ± 17.45.

ns, *, ** and *** indicate non-significant or significant at P ≤ 0.05, 0.01, 0.001, respectively.

ENHANCEMENT OF ABA ACTIVITY USING CYTOCHROME P450 INHIBITORS

Dale O. Wilson[1,2], Jr., Kimberly A. Falco[1], Rebecca M. Dickinson[1], F. Paul Silverman, Peter D. Petracek[1] and Daniel Leep[1]

[1]Valent BioSciences Corporation, 870 Technology Way, Libertyville, IL USA

[2]Corresponding author: dale.wilson@valentbiosciences.com

Abscisic acid (S-ABA) has recently been commercialized for agricultural uses by Valent BioSciences (Libertyville, IL). Despite improvements in manufacturing resulting in greater availability of S-ABA, application rates, and length of effectiveness have limited S-ABA uses. The discovery of S-ABA synergists would facilitate new uses of S-ABA in commodity agricultural crops. S-ABA level in the plant depends on a balance between synthesis and degradation. The major S-ABA catabolic pathway in plants is 8'-hydroxylation via the cytochrome P450 CYP707A family, although other degradation routes through hydroxylation are known. Many azole compounds that inhibit cytochrome P-450 enzymes are promiscuous, inhibiting other non-target P-450 enzymes. Plant growth regulators that inhibit gibberellin synthesis (CYP701a) may also suppress S-ABA catabolism. Likewise, many fungicides that are 14α-demethylase (CYP51) inhibitors might serve as S-ABA synergists. In addition, piperonyl butoxide (PBO), an insecticide synergist and inhibitor of insect cytochrome P-450 monooxygenases could be useful. Using a soybean time-to-germination assay as a model system, we tested many azole drugs with and without S-ABA. We also tested several methylene dioxyphenyl and linear furanocoumarin natural products as potential S-ABA synergists. In the cotton leaf transpiration system, we tested PBO, and found that it increased S-ABA activity. Several compounds, especially diniconazole and metconazole, appear promising as S-ABA synergists.

DEVELOPING A GROWING SEASON PHENOLOGY MODEL FOR PISTACHIO CULTIVARS

Cara Allan[1,2], Louise Ferguson[1]

[1]Plant Sciences Department, University of California, Hutchison Drive, Davis, California 95616 USA

[2]Corresponding author: callan@ucdavis.edu

With the rise in interest in systems modelling there is an opportunity for pistachio growers to take advantage of new technologies to enhance their production. Management decisions have been made in other industries by using phenology models as tools. For example, the pressure of the olive fly in olive production is only severe when the olive is 80 mm^3 in size and therefore tracking fruit development in a phenology model has shown spraying for the pest before the crop has accumulated 1200 heat units is unnecessary. Heat unit accumulation has been shown as a driver of development in fruit, especially well documented in peach and applied to almond production. We propose to expand the use of heat unit accumulation by characterizing nut growth as a function of heat units. Simultaneously, we will document stage development of the pistachio nut using biomarkers for the individual stages. By tracking this process stage development as a product of heat unit accumulation our research will be a tool in pest and disease control.

POSTHARVEST TREATMENT OF TOMATO WITH PGRS TO EXTEND FRUIT SHELF LIFE

Veria Alvarado[1,2,3] and Jerry Stoller[1]

[1]Stoller Enterprises, Inc., 4001 W Sam Houston Pkwy N, Houston, TX 77043 USA

[2]Institute for Plant Genomics and Biotechnology, Borlaug Center, Texas A&M University, College Station, Texas USA

Corresponding author: valvarado@neo.tamu.edu

A large portion of all fresh produce is lost worldwide after harvest. The main causes are physiological (wilting, shriveling, chilling injury, etc), pathological (decay due to fungi and bacteria) and physical (mechanical injury), being these causes in many instances interrelated, i.e., mechanical injury can lead to postharvest decay in many cases. Losses are estimated at 20-40% in developing countries and 10-15% in developed countries, depending on the crop. Just in the EU an estimated 4 billion EUR is lost due to postharvest losses and reduced quality of fruit. Current postharvest practices such as 1-MCP, a gaseous ethylene-binding inhibitor, do not produce consistent efficacy in delaying ripening. Postharvest applications in developing countries fail due to lack of sealed containers and time that 1-MCP takes to treat pallets. Therefore, a product applied in liquid form will be more suitable for postharvest treatments. We have developed a product named PRESERVE that consists of Calcium Chloride, Gibberellic Acid, and Salicylic Acid, which significantly delays ripening in tomato fruit stored either at 25 °C or 10 °C. Future plans include the application of PRESERVE in other climacteric fruits and the evaluation of pre harvest applications of PRESERVE.

INVOLVEMENT OF *ARF6* AND *ARF8* AUXIN RESPONSE FACTORS AND JASMONIC ACID IN TISSUE REUNION PROCESS OF INCISED *ARABIDOPSIS* INFLORESCENCE STEMS

Masashi Asahina[1,4], Weerasak Pitaksaringkarn[2], Keita Matsuoka[2], Miho Shimizu[2], Sumie Ishiguro[3], Emi Yumoto[1], Takao Yokota[1], Hisakazu Yamane[1], Shinobu Satoh[2]

[1]Department of Biosciences, Teikyo University

[2]Life and Environment Sciences, University of Tsukuba

[3]Graduate School of Bioagricultural Sciences, Nagoya University

[4]Corresponding author: asahina@nasu.bio.teikyo-u.ac.jp

Previously, we revealed that auxin and wound-inducible hormones contributed to the control of tissue reunion process in upper and lower part of incised stem by inducing the expression of *ANAC071* and *RAP2.6L*, respectively. The expressions of *ANAC071* and *RAP2.6L* were suppressed in incised stem of *arf6arf8* double mutant, but *RAP2.6L* expression was enhanced in non-incised stem of *arf6arf8*. The *arf6arf8* mutant showed inhibition of cell division in pith tissue after a week of incision, while no inhibition was observed in each single mutant and in the mutants for the other *ARFs*. We also found that some JA-related mutants showed incomplete healing process in tissue reunion.

QRT-PCR analyses showed that some of JA biosynthesis genes were up-regulated and the JA level was increased during the tissue-reunion process. In addition, expression of *DAD1* (DEFECTIVE IN ANTHER DEHISCENCE1) was suppressed in *arf6arf8*. From these results, we hypothesized that auxin signaling via ARF6/8 is essential for the expression of *ANAC071* and *RAP2.6L* in upper and lower part of incised stem, and the production of JA via induction of *DAD1* to induce *RAP2.6L* expression in tissue reunion process.

ACKNOWLEDGEMENT

This work was supported in part by JSPS-KAKENHI [Grant-in-Aid for Young Scientists B; 26840098 to M.A.], [Grant-in-Aid for Scientific Research on Innovative Areas; 24114006 to S.S.] and MEXT-supported Program for the Strategic Research Foundation at Private Universities [S1311014 to M.A.].

LITERATURE CITED

Asahina M, Azuma K, Pitaksaringkarn W, Yamazaki T, Mitsuda N, Ohme-Takagi M, Yamaguchi S, Kamiya Y, Okada K, Nishimura T, Koshiba T, Yokota T, Kamada H, Satoh S (2011). Spatially selective hormonal control of RAP2.6L and ANAC071 transcription factors involved in tissue reunion in Arabidopsis. Proc. Natl. Acad. Sci. USA 108: 16128-16132.

Pitaksaringkarn W, Ishiguro S, Asahina M, Satoh S (2014). ARF6 and ARF8 contribute to tissue reunion in incised Arabidopsis inflorescence stems. Plant Biotechnol. 31, 49-53.

FOLIAR APPLIED PACLOBUTRAZOL INFLUENCES PHYSIOLOGY AND MORPHOLOGY OF YOUNG PEACH TREES

Dennis E. Deyton[1,3], *C.G. Milne*[2] *and Carl E. Sams*[1]

[1]*The University of Tennessee, Department of Plant Sciences, Knoxville, TN, 37996-4561 USA*

[2]*Pellissippi State Technical Community College, Knoxville, TN 37932, USA*

[3] *Corresponding author: deytond@utk.edu*

INTRODUCTION

Paclobutrazol (PBZ) is a triazole plant growth regulator (PGR) that is a gibberellin biosynthesis inhibitor. Researchers evaluated it extensively during the 1980s and 1990s, and it was labeled for use on ornamental trees (Chorbadjian et al., 2011), floriculture (Runkle, 2011) and turf (Bigelow, 2012) in the USA. PBZ was evaluated for vegetative growth and fruiting of fruit trees in numerous trials (Miller and Swietlik, 1986; Yoshikawa et al., 1988; Greene, 1991) but was not labeled for fruit crops in the USA. It was reported that almost 80% of apple orchards in 2004 in the United Kingdom were treated with PGRs, and PBZ accounted for 54% of PGR usage (Crop Factsheet, 2007). PBZ is apparently not currently used on fruit in Europe (personal communication at the 2014 PGRSA meeting). The EU Pesticides database (Directorate General, 2008) lists PBZ maximum residue level of 0.5 mg/kg of for most fruit.

However, PBZ was withdrawn from inclusion of authorization for plant protection products as listed by the Commission Regulation in 2008 (EFSA, 2010). The formulation PBZ formation Payback® (Crop Care Australasia Pty Ltd, Murarrie Qld, AU) is labeled in Australia for application as a collar drench or through trickle irrigation to peach trees and as a foliar spray on apple trees (Crop Care Australasia, 2014). Layne et al. (2013) reported that peach trees grown in protective culture in China were sprayed with 750 $mg \cdot L^{-1}$ PBZ to stop vegetative growth and promote flower bud development. Recent research in China (Zhang et al., 2013) reported that spraying outdoor peach trees two times with no higher than 1500 $mg \cdot L^{-1}$ PBZ or greenhouse peach trees two times with 1000 $mg \cdot L^{-1}$ did not leave residues in mature peach fruit higher than the MRL allowed in fruit in Europe. PBZ is used primarily to reduce vegetative growth, but reports differ on its effect on aspects of tree physiology, such as net photosynthesis (Pn) and drought stress tolerance.

The objectives of this project were to determine the effects of PBZ concentration on photosynthesis, vegetative growth, and drought stress tolerance of young peach trees.

MATERIALS AND METHODS

Experiment 1. PBZ effect on tree growth and whole tree photosynthesis. One-year-old 'Redhaven' peach trees on 'Lovell' rootstock were planted in 19 L plastic pots in mid-May. Each tree was pruned to 21 cm above the graft union and then to a single shoot at ≈ two weeks before treatment. The plants were grown outdoors on black plastic with drip irrigation. Treatments of 0, 300, 600, 1200 or 2400 $mg \cdot L^{-1}$ PBZ (Cultar) were sprayed until runoff when the shoots had grown ≈ 10 cm. Treatments were emulsified with 0.1% (v/v) Tween 20. Pot and soil surfaces were covered with plastic to reduce spray contact. Treated trees were arranged in a randomized complete block (RCB) design with eight replications and two trees per plot. The trees were stored outdoors during the winter with protection to minimize cold damage. Trees were arranged in a RCB design during a second growing season, but with one tree per plot. Measurements were made at the end of both growing seasons of central shoot (leader) growth, number of nodes on the leader, number of lateral shoots and total shoot growth.

Whole tree photosynthesis was measured on five replications of two trees/plot the first growing season. Trees were moved into a greenhouse the day before measurement and then under HID metal halide lamps for one hour to allow acclimation before placing in a closed static assimilation chamber. The chamber sides were composed of transparent Plexiglas. The plants were exposed to 1000 $\mu mol \cdot m^{-2} \cdot s^{-1}$ PPF provided by HID metal halide lamps. Temperature inside the chamber was maintained at 27 °C ± 1 °C. Pn of trees from one replication were measured on a single day. Thus, Pn of five replications of trees were measured from 26 August until 30 August. At ten minute intervals, 60 cc gas samples were removed by syringe. Samples from the first replication were measured with an Anarad AR600 infra-red gas analyzer. Because of equipment difficulties, latter CO2 measurements were made with a Shimadzu gas chromatograph.

Experiment 2. PBZ effect on leaf photosynthesis and tree growth. An experiment was conducted the next year, again on one-year-old 'Redhaven'/'Lovell' peach trees planted in 19 L plastic pots in mid-May. The trees were grown similarly, the same treatments applied as described above. Treated trees were arranged in a RCB design with five replications of single tree plots. A new fully expanded leaf was tagged at treatment and referred to as "mature" leaf. Pn and stomatal conductance (gs) rates of a mature leaf and a new fully expanded leaf were measured outdoors on each tree. Pn was measured 16, 48 and 62 days after treatment (DAT) using an ADC open system infrared CO2 analyzer with a Parkinson leaf chamber. Data for young leaves measured 16 DAT were corrupted and not used. Stomatal conductance was measured with a LI-COR 1600 steady state porometer at 9, 23, 38 and 44 DAT. All measurements were made between 11:00 AM and 2:00 PM when photosynthetic photon flux was > 800 $\mu mol \cdot m^{-2} \cdot s^{-1}$. The trees were grown a second year, but bacterial spot severely damaged the foliage.

Experiment 3. PBZ effect on drought stressed trees. In a third experiment, one-year-old 'Redhaven'/'Lovell' peach trees were planted in 66 L pots the first week of June and grown in a greenhouse. The same treatments as described above were applied on 9 July and then arranged in a RCB design with four replications of five treatments and two tree plots. The trees/pots were saturated with water for three consecutive days (starting 1 September) prior to initiation of drought stress. All water supplies were eliminated from 4 September until 10 October to induce drought stress. A leaf from the midpoint of the leader (central shoot) of each tree had Pn measured with the ADC CO2 analyzer, and gsmeasured with the

LI-COR porometer at the start of the trial and then removed to measure water potential with a PMS 650 pressure chamber. Measurements were made throughout the trial, each time on the next highest leaf on the central shoot. Access tubes were placed in each pot and soil moisture measured with a Troxler 3220 neutron probe depth moisture gage that had been calibrated with gravimetric values. The trees were harvested 13 October, and measurements taken of leaf area and shoot elongation. The trees were partitioned into central and lateral shoots, leaves, and roots and the dry weights determined.

RESULTS

Experiment 1. PBZ effect on tree growth and whole tree photosynthesis. Foliar sprays of 300 to 2400 mg·L^{-1} PBZ reduced peach shoot elongation growth compared to the growth of control trees during the first growing season. Total shoot elongation was 13% to 35% less on PBZ treated trees ($P \leq 0.05$) than on untreated trees, with 2400 mg·L^{-1} PBZ causing the greatest reduction (Table 1). PBZ sprays reduced central shoot (leader) elongation by 4% to 16% ($P \leq 0.05$) in the year of treatment, compared to controls. The reduced growth of the peach leader was due to reduced internode length rather than reduced number of nodes. Reduction of peach leader growth was much less than the 30% to 46% reduction for 'Golden Delicious' apples in a concurrent trial (in press). The PBZ sprays reduced total lateral growth ($P \leq 0.05$) of the young peach trees by (18% to 55%), a greater effect than on the leader. The greater reduction of total lateral shoot growth than leader growth was due to reduced number of shoots as well as mean shoot elongation (data not shown) of PBZ treated trees. PZB did not significantly affect shoot elongation or number of branches the second growing season.

Foliar sprays of 300 to 2400 mg·L^{-1} PBZ did not affect whole peach tree Pn rates in this trial at ≈ 40 DAT. The lack of differences in whole plant Pn rates may have been due to lack of efficacy of the treatments, the equipment used, or the methodology used.

Experiment 2. PBZ effect on leaf photosynthesis and tree growth. Plant growth in this trial was measured as accumulated dry weights of leaders, shoots/branches and roots at the end of the second growing season. Dry matter weights were either too variable or sample size too small among treatments to differ significantly for most measurements (data not shown). The bacteria spot damage to leaves probably influenced results. It is noted that shoot/branch dry weights decreased linearly ($P \leq 0.05$) as PBZ concentration increased, with shoots/branches of trees treated with 2400 mg·L^{-1} PBZ weighing 46.2 g compared to shoot/branch of untreated trees weighting 72.7 g.

The results of PBZ concentrations on individual leaf Pn rates were very different from results of whole plant Pn measurements in the previous experiment. Pn rates of newly expanded leaves at 48 DAT increased linearly with PBZ concentration, with leaves from 2400 mg·L^{-1} PBZ treated trees having 53 % higher Pn rates than controls (Fig. 1A). At 62 DAT, the Pn rates of newly expanded leaves of PBZ treated trees were 6% to 17% higher ($P > 0.05$) than controls. Pn rates of mature leaves sprayed with ≥ 600 mg·L^{-1} PBZ were 4% to 27%, 12% to 17%, 8% to 13% higher than untreated mature leaves at 16, 48, and 62 DAT, respectively (Fig. 1B). However, Pn rates of PBZ treated mature leaves did not differ

($P > 0.05$) from controls on any one date. PBZ did not affect stomatal conductance of young leaves even though the gs rates of leaves on plants treated with ≥ 600 mg·L^{-1} were numerically higher than controls at 9, 23 and 38 DAT (Fig. 2A). Mature leaves present at treatment with ≥ 600 mg·L^{-1} PBZ had > 7 % higher gs rates than control leaves at 9 and 38 DAT, and > 80% higher at 23 DAT when gs rates were especially low (Fig. 2B). None of the treatment effects were significantly different on any one date.

Experiment 3. PBZ effect on drought stressed trees. All PBZ treated trees were 13% to 20 % shorter than controls in October at termination of the drought-stress trial, though not significantly different (Table 2). The number and average length of lateral shoots decreased linearly with increasing PBZ concentration, resulting in less total lateral shoot elongation ($P \leq 0.01$). Total shoot elongation was reduced by 29% to 50% by PBZ treatments. There were numerically 8% to 25% fewer leaves, but larger leaves on PBZ treated trees than on untreated trees.

The dry weights of central stems (without leaves) of PBZ treated trees were 38% to 49% less ($P \leq 0.01$) than central stems of control trees (Table 3). The dry matter accumulation into lateral stems and total stem growth was also reduced by PBZ treatments. Even though the dry matter in individual leaves increased linearly with PBZ concentration treatment (Table 3), the number of leaves per tree was variable (Table 2) and total dry weight of leaves was not affected. Total shoot dry weights were less for PBZ treated trees but root dry weights were not affected. Thus, root to shoot ratio was greater for PBZ treated trees.

A comparison of asymptotic lines of the relationship leaf Pn rates and soil moisture indicated that Pn rates of 2400 mg·L^{-1} PBZ treated trees were less than for leaves of other treatments when soil moisture was > 10% v/v (even though the Pn rates were numerically similar to rates from other treatments) (Fig. 3). Leaves from 2400 mg·L^{-1} PBZ trees (Fig. 3A) maintained higher Pn rates than trees treated with lower concentrations (1200 mg·L^{-1}, shown in Fig. 3B) or controls (Fig. 3C) as soil moisture declined from 5% to 0.5% (v/v). No differences in gs rates were found due to treatments.

Conclusions

These experiments demonstrated that foliar sprays of PBZ have vegetative growth controlling properties for peach trees during the current growing season. Although PBZ is a relatively stable gibberellin inhibitor, early season single-sprays of 300 to 2400 mg·L^{-1} had little effect on the vegetative growth of young trees the following growing season. This differs from a report (Arzani et al., 2009) that soil application of PBZ before anthesis reduced shoot elongation and pruning dry weights during the year of application and the following year. The foliar treatments reduced shoot elongation primarily by reducing internode length rather than number of nodes.

PBZ has been reported to reduce peach leaf size (Abou El-Khashab et al., 1997; Young, 1983) but leaves on trees treated with PBZ in Experiment 3 were larger than on controls. Total shoot dry weights were less for PBZ treated trees but root dry weights were not affected, thus resulting in increased root to shoot ratio (R/S). The effect of PBZ on R/S can affect nutrient uptake (Rieger and Scalabrelli, 1990) and potentially stress tolerance.

The effects of PBZ on fruit tree photosynthesis have been variable in other reported trials. DeJong and Doyle (1984) found no apparent decrease in Pn of nectarine leaves due to PBZ treatment. Rieger and Scalabrelli (1990) reported that PBZ in nutrient solutions did not affect Pn rates of young leaves of 'Nemaguard' rootstocks at 7, 21, or 28 days of treatment. Wieland and Wample (1985) reported that PBZ applied as a soil drench or directly to apple stems did not affect Pn rates of mid-shoot apple leaves. However Huang et al. (1995) reported that soil-applied PBZ increased apple spur leaf Pn rates and contributed the increase Pn partly to higher light intensity in the canopy.

The results of foliar applications of PBZ on peach tree photosynthesis varied in each of our experiments. Foliar applications of PBZ had no effect on whole plant Pn rates in our initial experiment. The lack of effect on whole plant photosynthesis may have been due to our methodology or simply due to lack of treatment efficacy. However in experiment 2, Pn rates of mature leaves sprayed with ≥ 600 mg·L^{-1} PBZ were numerically higher than control leaves at 16, 48, and 62 DAT. Pn rates of young fully expanded leaves on PBZ treated trees were significantly higher at 48 DAT and numerically higher at 62 DAT than control leaves. However in experiment 3, Pn rates of leaves on PBZ treated trees were numerically similar to control leaves when soil moisture was >10% v/v. Thus, PBZ may increase photosynthesis in some cases but these trials do not show consistent trends.

Foliar sprays of PBZ appeared to impart some drought resistance to young peach trees, maintaining higher Pn rates when soil moisture declined below 5% v/v (Experiment 3). Soil applied PBZ has been reported to reduce drought stress effects on transpiration, growth and leaf water potentials of cherry rootstocks (Asamoah and Atkinson, 1985). Foliar sprays of PBZ have been reported to improve water-salinity tolerance of rooted 'Nemaguard' peach cuttings (Abou El-Khashab et al., 1997) and of strawberry plants (Jamalian et al., 2008). Marshall et al. (2000) found that PBZ treated tree seedlings survived drought treatments that killed untreated trees.

LITERATURE CITED

Abou El-Khashab, A.M., A.F. El-Sammak, A.A. Rlaidy, M.I. Salama and M. Rieger. 1997. Paclobutrazol reduces some negative effects of salt stress in peach. J. Amer. Soc. Hort. Sci. 122:43-46.

Arzani, K., F. Bahadori, and S. Piri. 2009. Paclobutrazol reduces vegetative growth and enhances flowering and fruiting of mature 'J.H. Hale' and 'Red Skin' peach trees. Hort. Environ. Biotechnol. 50:84-93.

Asamoah, T.E.O. and D. Atkinson. 1985. The effects of (2RS, 3RS)-l-(4-chlorophenyl)-4, 4-dimethyl-2-(1 H-1,2,4 triazol-l-yl)pentan-3-ol (Paclobutrazol:PP3 33) and root pruning on the growth, water use and response to drought of Colt cherry rootstocks. Plant Growth Regulation 3:37-45.

Bigelow, C.A. 2012. Plant growth regulators in bentgrass turf areas. USGA GreenSection Record. Vol. 50(8), 13 April 2012, http://gsrpdf.lib.msu.edu/ticpdf.py?file=/article/bigelow-plant-4-13-12.pdf (retrieved 20 June 2013).

Chorbadjian, R.A., P. Bonello, and D.A. Herms. 2011. Effect of the growth regulator paclobutrazol and fertilization on defensive chemistry and herbivore resistance of Austrian pine (*Pinus nigra*) and paper birch (*Betula papyrifera*). Intl. Soc. of Arboriculture. Arboriculture & Urban Forestry 37:278-287.

Crop Factsheet. 2007. Pesticides News 76. Apples – conventional, IPM and organic. http://www.pan-k.org/pestnews/Issue/pn76/pn76 p18-21.pdf (retrieved 20 June 2013).

Crop Care Australasia. 2014. Payback® Plant Growth Regulator. Pty Ltd, Unit 15/16 Metroplex Avenue, Murarrie Qld 4172. http://m.cropcare.com.au/assets/874/1/PaybackRLP592370713.pdf (retrieved 31 Feb., 2014)

DeJong, T.M. and J.F. Doyle. 1984. Leaf gas exchange and growth responses of mature 'Fantasia' nectarine trees to paclobutrazol. J. Amer. Soc. Hort. Sci. 109:878-88.

Directorate General. 2008. EU Pesticides database. From Reg. (EC) No 149/2008. http://ec.europa.eu/sanco_pesticides/public/?event=substance.historic (retrieved 7 August 2014).

EFSA. 2010. Conclusion on the peer review of the pesticide risk assessment of the active substance paclobutrazol. European Food Safety Authority, EFSA Journal 8(11):1876 (60 pp.).

Greene, D.W. 1991. Reduced rates and multiple sprays of paclobutrazol control growth and improve fruit quality of 'Delicious' apples. J. Amer. Soc. Sci. 116:807-812.

Huang, W.D., T. Shen, Z.H. Han, and S. Liu. 1995. Influence of paclobutrazol on photosynthesis rate and dry matter partitioning in the apple tree. J. Plant Nutr. 18:901–910.

Jamalian, S., A. Tehranifar, E. Tafazoli, S. Eshghi, and G.H. Davarynejad. 2008. Paclobutrazol application ameliorates the negative effect of salt stress on reproductive growth, yield, and fruit quality of strawberry plants. Hort. Environ. Biotechnol. 49:1-6.

Layne, D.R., Z. Wang, and L. Niu. 2013. Protected cultivation of peach and nectarine in China - industry observations and assessments. J. Amer. Pomol. Soc. 67:18-28.

Marshall, J.G., R.G. Rutledge, E. Blumwald, and E.B. Dumbroff. 2000. Reduction in turgid water volume in jack pine, white spruce and black spruce in response to drought and paclobutrazol. Tree Physiol. 20:701-707.

Miller, S.S. and D. Swietlik. 1986. Growth and fruiting response of deciduous fruit trees treated with paclobutrazol. Acta Hort. 179:563-566.

Rieger, M. and G. Scalabrelli. 1990. Paclobutrazol, root growth, hydraulic conductivity, and nutrient uptake of 'Nemaguard' peach. HortScience 25:95-98.

Runkle, E. 2011. A new paclobutrazol. http://www.flor.hrt.msu.edu/assets/Uploads/ Paclobutrazol.pdf (retrieved 20 June 2013). Yoshikawa, F.T., G.C. Martin, J.H. LaRue.

1988. Chemical growth regulator for peaches. California Agr., July-August, pp 19-20.

Young, R.S. 1983. Peach growth response from PP333 (paclobutrazol). 10th Annu. Mtg. of the Plant Growth Regulat. Soc. of Amer., pp 192-195.

Zhang W., J. Chen, J. Fang and M. Lin. 2013. Evaluation on residue amount of paclobutrazol in peach fruits. Acta Hort. Sinica, 40:1369-1374. http://www.ahs.ac.cn/EN/abstract/abstract4114.shtml# (retrieved 6 June 2014).

Table 1. First and second year shot growth of paclobutrazol treated one-year-old 'Redhaven' peach trees.

	First growing season				Second growing season		
	Central shoot		Total lateral shoots (cm)	Total shoots (cm)	Central shoot (cm)	Total latral shoots (cm)	Total shoots (cm)
PBZ[z] (mg·L⁻¹)	Length (cm)	Internodes (cm)					
0	41.5	1.82	42.6	84.1	74.4	509.3	583.7
300	38.1	1.62	34.9	73.0	66.6	398.8	465.4
600	39.7	1.66	30.8	70.5	67.2	527.4	594.6
1200	38.6	1.69	24.3	62.9	66.7	497.8	564.5
2400	34.8	1.68	19.3	54.1	70.2	458.3	528.5
Significance[y]	*	*	*	*	ns	ns	ns

[z] PBZ = Paclobutrazol, trees sprayed when shoots had grown approximately 10 cm.
[y] ns, *, non-significant, significant at $P \leq 0.05$.

Table 2. Growth of drought stressed one-year-old paclobutrazol treated 'Redhaven' peach trees.

	Shoot elongation[z]				Leaves		
PBZ ($mg \cdot L^{-1}$)[y]	Central shoot (cm)[x]	Total lateral shoots (cm)	Total shoots (cm)	Lateral shoots (no./ plant)	(no./tree)	(cm^2/ leaf)	(cm^2/ tree)
0	107.0	313.2	420.2	11.8	235	22.4	5253
300	94.4	111.9	206.3	9.2	199	32.4	6431
600	85.2	214.5	299.7	9.2	217	25.7	5564
1200	93.8	107.8	201.6	6.3	176	30.4	5360
2400	91.8	119.2	211.0	5.5	190	29.3	5552
Significance[x]	ns	*	*	*	ns	ns	ns

[z] Growth of trees from early June until 10 Oct. Drought stress imposed from 4 Sept. until 10 Oct. Shoots include current season stems and leaves.

[y] PBZ = Paclobutrazol, foliar treatments applied on 9 July.

[x] ns, *, nonsignificant, significant $P \leq 0.05$.

Table 3. Dry matter partitioning of drought stresses one-year-old paclobutrazol treated 'Redhaven' peach trees.[z]

	Stem			Leaves					
PBZ ($mg \cdot L^{-1}$)[y]	Central stem (g)	Total lateral stems (g/ plant)	Total stems (g/ plant)	(g/leaf)	(g/plant)	Shoots[x] (g)	Roots (g)	Plant (g)	Root/ shoot ratio
0	19.9	8.1	28.0	0.084	19.8	47.8	55.9	103.7	1.17
300	11.7	2.7	14.4	0.094	18.7	33.1	58.4	91.5	1.76
600	12.2	6.0	18.2	0.093	20.2	38.4	65.0	103.4	1.69
1200	10.8	3.6	14.4	0.102	18.0	32.4	52.7	85.1	1.63
2400	10.2	3.4	13.6	0.101	19.2	32.8	58.0	90.8	1.77
Significance[w]	**	*	*	*	ns	*	ns	ns	*

[z] Trees planted early June, exposed to drought stress from 4 Sept. until 10 Oct.

[y] PBZ = Paclobutrazol, trees sprayed on 9 July.

[x] Shoots = stem plus foliage.

[w] ns, *, ** nonsignificant, significant at $P \leq 0.05$ and $P \leq 0.01$, respectively.

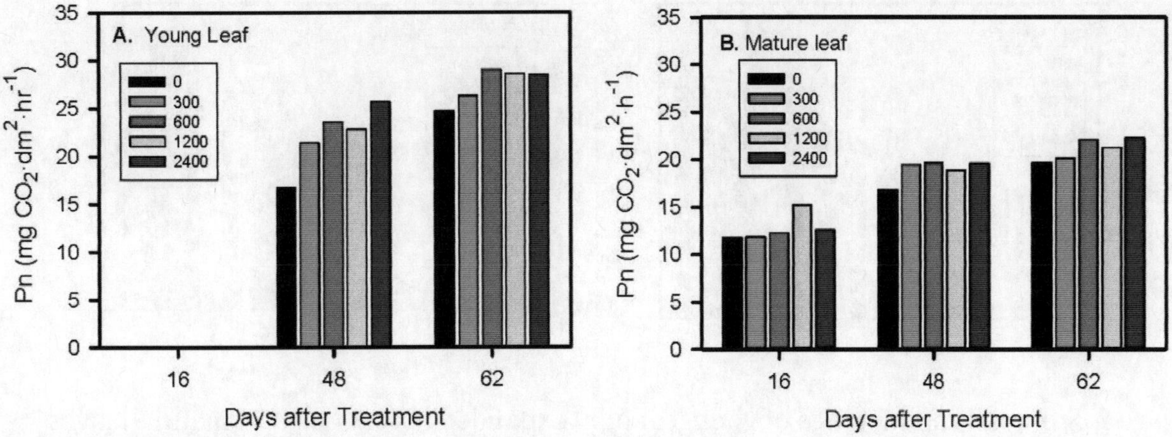

Figure 1. Net photosynthesis rates (Pn) of youngest fully expanded leaves (A.) and mature leaves (B.) of potted 'Redhaven' peach trees during the first season after application of 0 to 2400 mg·L^{-1} paclobutrazol when terminal growth was approximately 10 cm.

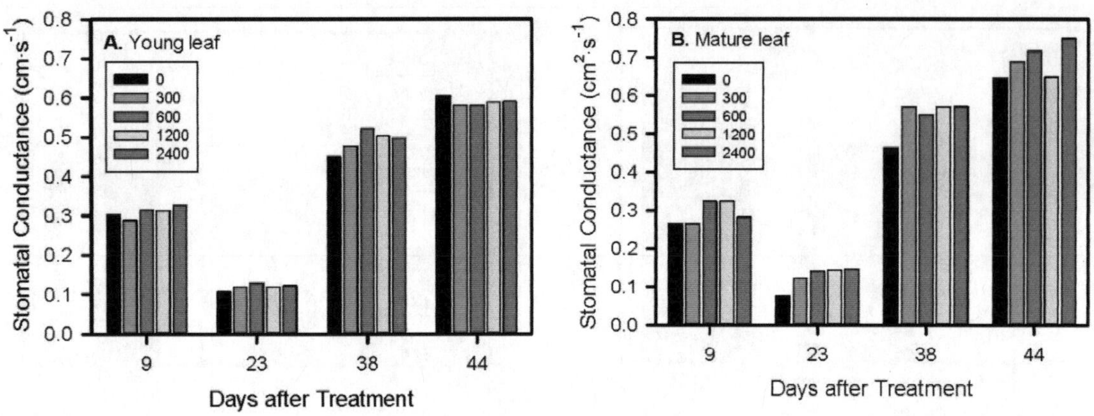

Figure 2. Stomatal conductance of youngest fully expanded leaves (A.) and mature leaves (B.) of potted 'Redhaven' peach trees during the first season after application of 0 to 2400 $mg{\cdot}L^{-1}$ paclobutrazol when terminal growth was approximately 10 cm.

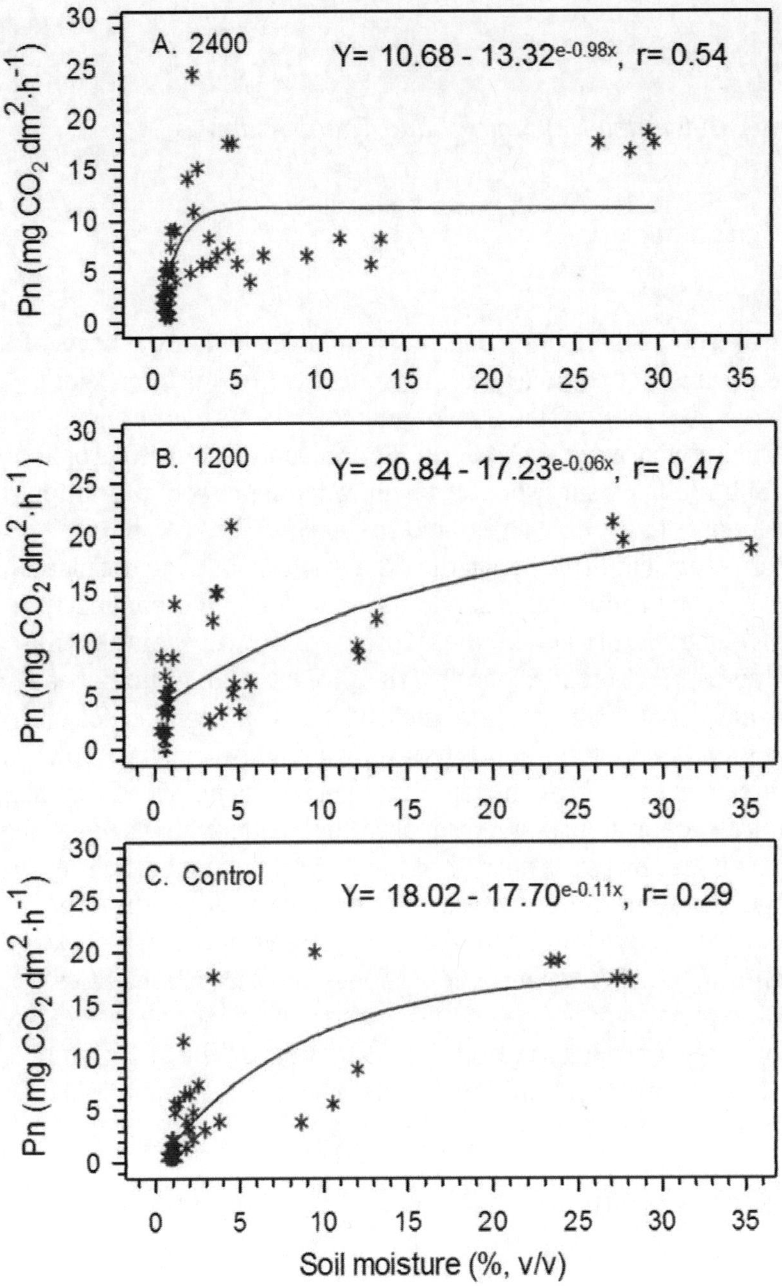

Figure 3. Net photosynthetic rates (Pn) of leaves of potted 'Redhaven' peach trees at different soil moistures after treatment with 2400 (A.) or 1200 mg·L^{-1} paclobutrazol (B.) or left unsprayed (C.) on 9 July and water supplies eliminated from 4 Sept. until 10 Oct. to induce drought stress.

THE STRIGOLACTONE MIMIC "DEBRANONES" IS APPLICABLE TO MANY ASPECTS OF AGRICULTURAL ISSUES

Kosuke Fukui[1,2], Daichi Yamagami[1], Hidemitsu Nakamura[1] and Tadao Asami[1]

[1]*The Univ. of Tokyo, Japan*

[2]*Corresponding author: fukui@pgr1.ch.a.u-tokyo.ac.jp*

Strigolactones (SLs) are recognized as a novel plant hormone, being involved in many aspects of plant growth and development. Meanwhile, SLs behave as a rhizosphere communication agent, which stimulates seed germination of root parasitic plants belonging to *Orobanchaceae* and induces hyphal branching of arbuscular mycorrhizal (AM) fungi. Plant hormonal funcitons of SLs are recently being uncovered well. SLs inhibit axillary shoot branching, influence root formation, and positively regulate the plant responses to drought stress and so on. Therefore, SLs have potential as a plant growth regulator for agricultural use. On the other hand, root parasitic plants such as *Striga* cause severe damage to crop production in sub-Saharan Africa. In this context, the research of SLs should be beneficial for agriculture and biomass resources. However, natural SLs were poorly available for academic and agricultural use because they are less stable and not easily synthesized. So, the development of easily obtainable and stable SL agonists has significance with regard to above issues. Recently we found novel compounds that can be readily prepared from phenolic compounds and exert SL action in shoot branching inhibition assay using rice SL deficient mutant. We called this type of chemicals "debranones". We prepared several derivatives of debranones and evaluated their SL activities by two typical SL assays, axillary shoot branching inhibition assay of plants and seed germination stimulation assay of root parasitic plants. As a result of structure-activity relationship study, we found that there are close relationships between a variety and positions of substituents on the phenyl ring of debranones and their SL activities. Through this research we found unique compounds. One compound showed strong shoot branching inhibition activity but not strong germination stimulation activity. On the other hand, another compound showed weak shoot branching inhibition activity but strong germination stimulation activity.

FLAGELLIN GLYCANS OF *ACIDOVORAX AVENAE* RICE VIRULENT K1 STRAIN REGULATE INDUCTION OF RICE IMMUNE RESPONSES

Hiroyuki Hirai[1,2], *Yukihiko Nakagawa*[1], *Takehito Furukawa*[1] *and Fang-Sik Che*[1]

[1]*Graduate School of Biosciences, Nagahama Institute of Bio-Science and Technology*

[2]*Corresponding author: h_hirai@nagahama-i-bio.ac.jp*

Acidovorax avenae is a gram-negative plant pathogenic bacterium. The flagellin from *A. avenae* rice avirulent N1141 strain induced several rice immune responses including H2O2 generation, while the flagellin from rice virulent K1 strain did not. To determine whether differences in the amino acid sequences between the N1141 and K1 flagellins primarily cause the specific immune responses in rice cells, His-tagged N1141 and K1 flagellins were produced in *E. coli.* His-tagged recombinant K1 or N1141 flagellin equally induced H2O2 generation when cultured rice cells were treated with these flagellins, suggesting that different amino acid residues of 14 sites between N1141 and K1 flagellins did not involve in the specific induction of immune responses (Fig. 1). We assumed that post-translational modifications existed flagellins of *A. avenae*, and flagellin glycans of *A. avenae* involved in the specific induction of immune responses in rice. Mass spectral analysis and glycan analysis showed that total 1,600 and 2,150 glycans were present on the N1141 and K1 flagellins, respectively (Table 1). Deficient mutants (NÄ*Fgt* and KÄ*Fgt*) of putative flagellin glycosyltransferase gene in N1141 and K1 strains that existed in flagella operon were produced by homologous recombination. Mass spectral analysis and glycoprotein staining analysis revealed that flagellin of NÄ*Fgt* and KÄ*Fgt* mutants did not glycosylate (Table 1). A deglycosylated K1 flagellin induced H2O2 generation in the same manner as N1141 flagellin (Fig. 2). Tryptic peptide mappings with reverse-phase HPLC and mass spectrometer revealed that glycans were attached to four amino acid residues (^{178}Ser, ^{183}Ser, ^{212}SER and ^{351}Thr) in K1 flagellin and three amino acid residues (^{178}Ser, ^{183}Ser and ^{351}Thr) in N1141 flagellin. Mutants of N1141 and K1 strain in which glycan-attached amino acid residues were substituted alanine amino acid residues were generated. Mass spectral analysis of flagellins in N1141 and K1 mutants showed that the molecular weight of each glycan in N1141 and K1 flagellins were 540 and 551, respectively. Among mutant K1 flagellins in which each glycan-attached amino acid residue was changed to alanine, ^{178}Ser/Ala and ^{183}Ser/Ala K1 flagellin induced a strong immune response in cultured rice cells, indicating that the glycans at ^{178}Ser and ^{183}Ser in K1 flagellin prevent epitope recognition in rice (Fig. 3).

Table 1. Molecular masses of intact flagellin (Ala^{2}-Arg^{492}).

	$[M + H]^+$		
	Measured molecular mass of flagellin	Calculated molecular mass of flagellin	Measured - Calculated
N1141			
Wild type	50,790	49,257	1,533
NΔ*Fgt*	49,280	49,257	23
K1			
Wild type	51,248	49,112	2,136
KΔ*Fgt*	49,090	49,112	-22

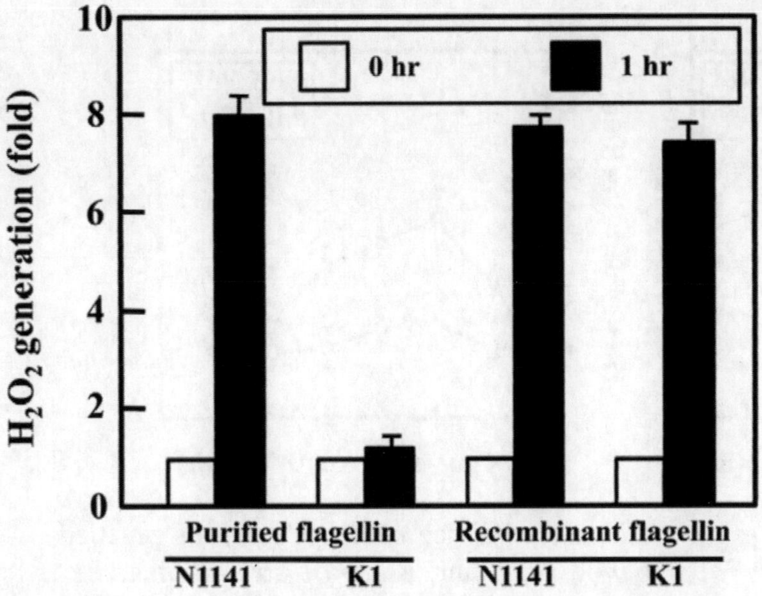

Figure 1. Induction of immune responses in cultured rice cells by flagellin from *A. avenae.* Time course of H2O2 generation in cultured rice cells that were treated with flagellin purified from the avirulent N1141 strain or virulent K1 strain. The y axis represents the -fold change in H2O2 in cultured rice cells relative to the levels before flagellin treatment. White columns, 0 h after treatment; black columns, 1 h after treatment.

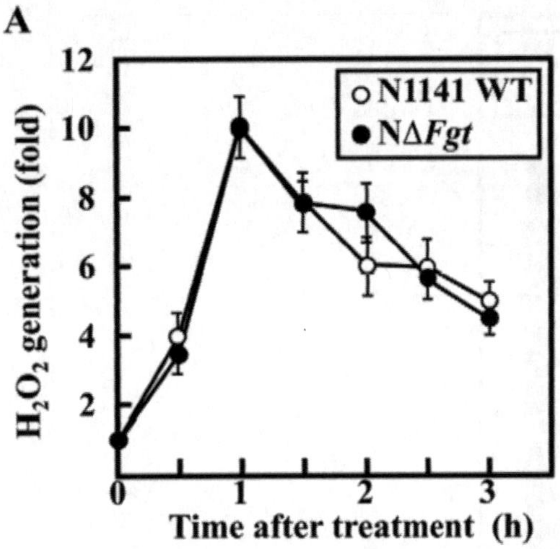

Figure 2. Induction of immune responses in cultured rice cells by flagellins purified from N1141 wild type, N1141Ä*Fgt*, K1 wild type, and K1Ä*Fgt* strains of *A. avenae*. A, time course of H2O2 generation in cultured rice cells that were treated with flagellin purified from the N1141 wild-type (open circles) or N1141Ä*Fgt* (solid circles) strain. B, time course of H2O2 generation in cultured rice cells that were treated with flagellin purified from the K1 wild-type strain (open circles) or K1Ä*Fgt* (solid circles) strain. H2O2 was detected using a luminol chemiluminescence assay.

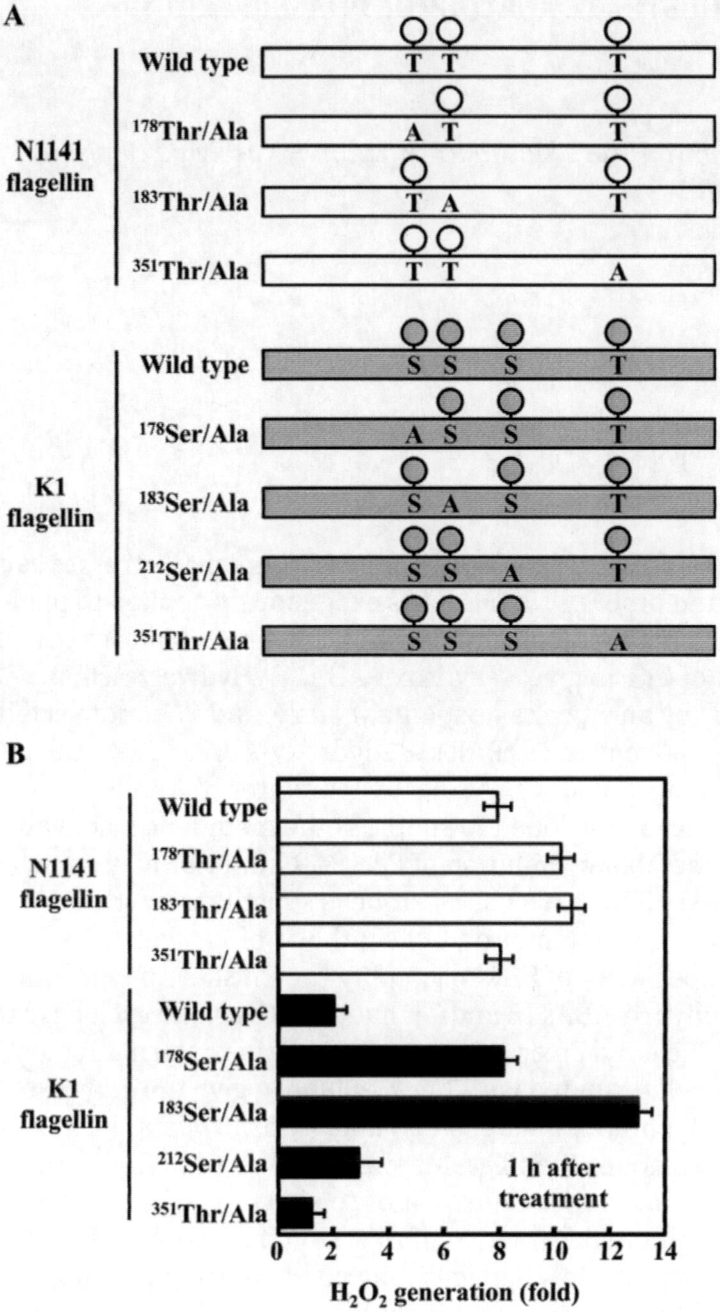

Figure 3. Induction of immune responses in cultured rice cells by several mutant flagellins. A. schematic diagram of N1141 and K1 mutant flagellins. White columns, N1141 flagellin; white circles, N1141 flagellin glycan; gray columns, K1 flagellin; gray circles, K1 flagellin glycan. B. H2O2 generation in cultured rice cells that were treated with several mutant flagellins.

SYNTHESIS OF FLUORINATED SUBSTRATE ANALOGS OF PHASEIC ACID 4'-REDUCTASE

Kenta Usami[1,5], Yusuke Fujii[1], Nobuhiro Hirai[1], Masaharu Mizutani[2], Toshiyuki Ohnishi[3], Yasushi Todoroki[3] and Satoru Kondo[4]

[1]*Kyoto University*

[2]*Kobe University*

[3]*Shizuoka University,*

[4]*Chiba University*

[5]*Corresponding author: hirai@kais.kyoto-u.ac.jp*

A plant hormone, abscisic acid (ABA, 1) protects plants from environmental stresses such as water stress, low temperature and others. When ABA is exogenously applied to plants, its effect, however, is quickly lost by the metabolic inactivation. The metabolic inactivation of ABA begins from hydroxylation of C-8' to give 8'-hydroxy-ABA. 8'-Hydroxy-ABA is metabolized to phaseic acid (PA, 2) by an enzyme or spontaneously, and PA is converted to dihydrophaseic acid (DPA, 3) by PA 4'-reductase (PAR). 8'-hydroxy-ABA and PA still have a low biological activity, but DPA is almost inactive. A selective inhibitor of ABA 8'-hydroxylase, abscinazole-E2B, has been developed to suppress the metabolic inactivation of ABA. Another target is PAR for metabolic inhibition of PA. PAR inhibitor may show long lasting of ABA activity. An analog (4) of PA in which 4'-ketone is substituted with a difluoromethylene groups may act as a PAR inhibitor, but synthesis of PA skeleton is not easy. We examined the substrate specificity of PAR to simplify the PA skeleton for PAR inhibitor, and revealed that 2',3'-dihydro-ABAs (6 and 7) that have the same cyclohexanone ring as PA were reduced by PAR. The ethyl ester of 6 that has a stable conformation with an axial side chain similar to PA was fluorinated with Deoxo-Fluor to give the ethyl esters of 8 and 9 at a ratio of 1:2. The PAR inhibitory activity of 8 and 9 was low, but this result suggested that PA having a saturated double bond at C-2' was fluorinated at C-4' with Deoxo-Fluor. Actually, PA was fluorinated with Deoxo-Fluor to give (1'*S*)-1',4',4'-trifluoro-PA (F_3-PA, 5) at a low yield. F_3-PA inhibited formation of DPA from PA by PAR with an inhibitory ratio of 70%, but F_3-PA did not show a long lasting effect of ABA in a rice seedling elongation assay. 4',4'-Difluoro-PA may be more effective inhibitor of PAR than F_3-PA.

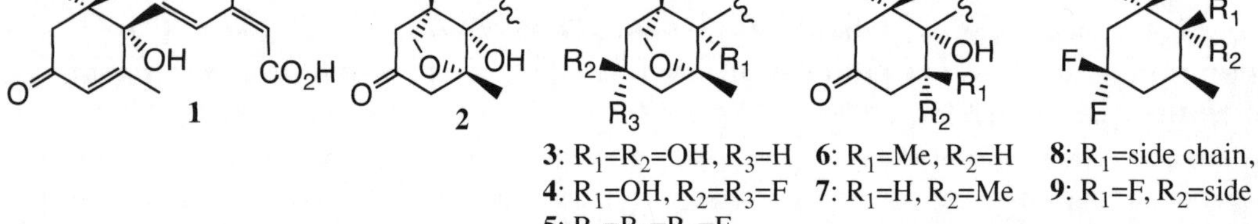

3: R_1=R_2=OH, R_3=H **6**: R_1=Me, R_2=H **8**: R_1=side chain, R_2=F
4: R_1=OH, R_2=R_3=F **7**: R_1=H, R_2=Me **9**: R_1=F, R_2=side chain
5: R_1=R_2=R_3=F

PRELIMINARY STUDY ON PGR REGULATION OF BIOMASS PRODUCTION IN AN ENERGY GRASS *TRIARRHENA LUTARIORIPARIA*

Yuan Xiao[1], Zhiwei Xiang[1], Qingyang Liu[1], Yahui-Hong[1], Zhigang Huang[1,2] and Langtao Xiao[1]

[1]Hunan Provincial Key Laboratory of Phytohormones and Growth Development, College of Bioscience and Biotechnology, Hunan Agricultural University, Changsha, China

[2]Corresponding author: klphzg@163.com

Triarrhena lutarioriparia L. Liu is a high biomass yielding fibrous perennial C4 plant native to the everglade of several inland lakes in southern China, thus it is regarded as a potential new energy grass. In order to promote its exploitation and application, plant growth regulators of gibberellin(GA3) and brassinolide (BL) were foliar sprayed respectively in pot and plot experiments. The effects of the above PGRs on growth, biomass yield were studied. The plant height, stem diameter and dry weight were determined after different PGR treatments. The preliminary results showed that GA treatment and BR treatment significantly increased the plant height and stem diameter respectively. The combination treatment of GA plus BR increased significantly the biomass (dry weight) by moderately increasing the plant height and stem diameter. Larger scale experiments based on airplane spray is also in progress.

ACKNOWLEDGEMENT

Supported by National Natural Science Foundation of China, Grant 91317312, Scientific Research Fund of Hunan Provincial Education Department, Grants 12K060 and 12K061)

FUNCTION OF STRIGOLACTONE AS A CHEMICAL REGULATOR OF PHOSPHATE STARVATION SIGNALING

Shinsaku Ito[1,5], Tomoko Nozoye[1], Kosuke Fukui[1], Hiromi Nakanishi[1], Naoko K. Nishizawa[1], Ken Ito[2], Shunsuke Yajima[2], Tadao Asami[2], Eriko Sasaki[3], Gregor Mendel[3], Misaki Imai[4], Yuh Shiwa[4], Mari Shibata-Hatta[4], Taichiro Ishige[4]

[1]The University of Tokyo, Japan

[2]Department of Bioscience, Tokyo University of Agriculture, Japan

[3]Institute of Molecular Plant Biology, Vienna, Austria

[4]Genome Research Center, Tokyo University of Agriculture, Japan

[5]Corresponding author: s4ito@nodai.ac.jp

Phosphate is an essential macronutrient in plant growth and development; however, the concentration of inorganic phosphate (Pi) in soil is often suboptimal for crop performance. Accordingly, plants have developed physiological strategies to adapt to low Pi availability. Here, we report that Pi starvation responses in *Arabidopsis* are partially dependent on the strigolactone (SL) signaling pathway. SL treatment induced root hair elongation, anthocyanin accumulation, activation of acid phosphatase, and reduced plant weight, which are characteristic responses to phosphate starvation. Furthermore, the expression profile of SL-response genes correlated with the expression of genes induced by Pi starvation. These results suggest a potential overlap between SL signaling and Pi starvation signaling pathways in plants.

STRUCTURE DETERMINATION AND FUNCTION ANALYSIS OF ACYL SPERMIDINES IN RICE

Junya Iwakawa[1,3], Masashi Hikosaka[1], Hidemitsu Nakamura[1], Tadao Asami[1], Satoru Maeda[2], Masaki Mori[2]

[1]University of Tokyo

[2]National Institute of Agrobiological Sciences

[3]Corresponding author: iwakawa@pgr1.ch.a.u-tokyo.ac.jp

Spermidine (Spd) is a well-known plant polyamine involved in several plant physiologies such as senescence and disease resistance responses and so on. Its N-hexanoylated derivative was reported to accumulate progressively during senescence in pea and therefore acylated Spds are expected to play roles in plants, but their physiological roles are not really investigated. Previously we found that several acylSpds accumulated in Spl18, a disease resistance mutant identified from rice activation-tagging lines and on the basis of this result we anticipated that acylSpds could have functions in plant immune system. Recently we demonstrated that N^4-(12-hydroxylauroyl)spermidine (4N 12HydLau) has similar structure to natural acylSpds accumulated in Spl18 and confers resistance to disease in rice. Thus we attempted to identify the structures of these natural acylSpds accumulated in rice and thereafter we examined the effect of them on plant disease resistance.

EXPERIMENT AND RESULT

Firstly, we synthesized several 4N 12HydLau isomers from Spd and hydroxylauric acids by using EDC as a condensing agent. We compared their physiochemical characters with those of the acylSpds extracted from the Spl18 mutant rice by LC-MS/MS. As a result, we found that three acylSpds, in which positions being hydroxylated are different from the position of hydroxylation in 4N 12HydLau, accumulate in the Spl18 mutant. Next, we examined the activities of these three natural acylSpds. It was suggested that these natural acylSpds treatment conferred resistance to Magnaporthe oryzae, which causes rice blast disease, and induced expression of pathogenesis-related genes, such as PBZ1 and OsPR1b.

DISCUSSION

Above results suggest that natural acylSpds should be potential plant activators, which induce plant disease resistance. However, the structures of acylSpds are largely different from those of pre-existing plant activators, such as probenazole and acibenzolar-S-methyl. Therefore, acylSpds may be plant activators with a novel mechanism of action. We are now attempting to elucidate the mechanisms that underlie the activation of plant immune systems by acylSpds.

NOVEL AUXIN-RESPONSIVE GENES OF ARABIDOPSIS ARE DISCOVERED USING MULTIPLE AUXIN INHIBITORS

Yusuke Kakei[1,4], Yosuke Ishida[1], Yukihisa Shimada[1], Ken-ichiro Hayashi[2] and Tadao Asami[3]

[1]Yokohama City University, Japan

[2]Okayama University of Science, Japan

[3]University of Tokyo, Japan

[4]Corresponding author: kakei@yokohama-cu.ac.jp

Although molecular studies of auxin-regulated gene expression have been conducted, these studies are mainly based on exogenous application of excessive auxin to plants. Recently, indole-3-pyruvate (IPyA) pathway was suggested as one of major biosynthesis pathway for Indole-3-acetic acid (IAA) in Arabidopsis. In this pathway, tryptophan aminotransferase (TAA1) converts Tryptophan to IPyA, and subsequently flavin-dependent monooygenase, YUCCA catalyzes the conversion from IPyA to IAA, the most important member of auxin family. We developed the first auxin-biosynthesis inhibitor aminooxyphenylpropionic acid (AOPP) that inhibits TAA1 and another inhibitor 4-phenoxyphenylboronic acid (PPBo) targeting YUCCA. We treated *Arabidopsis* seedlings with novel inhibitors for auxin biosynthesis and signaling inhibitor PEO-IAA. In addition to genes that have already been reported as auxin-responsive, we identified several gene families as novel auxin-responsive genes. Cell-wall-related genes, such as *xyloglucan endotransglycosylase hydrolases*, were regulated only by inhibitor treatments. Auxin-biosynthesis genes, *YUCCAs*, were up-regulated in response to auxin inhibitors. The results suggest that the genes regulated by the inhibitors are novel auxin-responsive genes, and that these genes are regulated by lower levels of auxin concentration.

THE PHYTOALEXIN MOMILACTONE BIOSYNTHESIS IN THE MOSS *HYPNUM PLUMAEFORME*

Honoka Kimura[1], Sho Miyazaki[1,2], Manami Shimane[1,5], Ryosuke Kainuma[1], Tomoko Ando[1], Masahiro Natsume[1], Ken-ichiro Hayashi[3], Hiroshi Nozaki[3], Kazunori Okada[4] and Hiroshi Kawaide[4]

[1]Institute of Agriculture, Tokyo University of Agriculture & Technology, Tokyo, Japan

[2]Department of Applied Biological Chemistry, The University of Tokyo, Tokyo, Japan

[3]Faculty of Science, Okayama University of Science

[4]Biotechnology Research Center, The University of Tokyo

[5]Corresponding author: 50014534012@st.tuat.ac.jp

Plants have some defense systems for many environmental stresses such as infection by pathogens, UV irradiation and exposure to heavy metals. The production of phytoalexins that induce growth inhibition of microorganisms and plants is one of the defense systems in plants. In rice, production of many types of phytoalexins is induced when the rice leaves and suspension-cultured cells are exposed to UV irradiation and treated with heavy metals, respectively. Momolactones are well known as rice diterpenoid phytoalexins and biosynthesized from GGDP *via syn*-copalyl diphosphete and 9βH-pimara-7,15-diene. Recently, our research group demonstrated that the moss *Hypnum plumaeforme* produces momilactoneA and B. We are interested in the biosynthesis and physiological roles of momilactones in *Hypnum* moss since rice and mosses are evolutionally different species. In this research, we cloned cDNAs encoding diterpene cyclases involved in the momilactone biosynthesis and *ent*-kaurene biosynthesis, the precursor of gibberellin plant hormone in flowering plants.

The cDNA library was prepared from gametophores (the aerial parts) of the *Hypnum* moss and homology-based PCR experiments successfully cloned a possible cDNA encoding a bifunctional diterpene cyclase (DTC, HpDTC1). We also analyzed all expression genes in the gametophores by a next-generation sequencer (Illumina Highseq 2000). Another candidate cDNA encoding a DTC was found from RNA-seq data and cloned from the library (HpDTC2). Bacterial expression and function analysis of both clones result that recombinant HpDTC1 and HpDTC2 enzymes were determined as *ent*-pimara-9,15-diene synthase and 9βH-pimara-7,15-diene synthase, respectively. Miyazaki *et al.* showed that *ent*-kaurene biosynthetic gene is expressed during protonemal growth in a model moss, *Physcomitrella patens.* GC-MS analysis of diterpene hydrocarbons extracted from both gametophore and protonemal cells of the *Hypmun* moss indicated that *ent*-kaurene was detected from only the extracts of the protonemal cells as the single hydrocarbon product. Based on the results of GC-MS analysis, we cloned a new DTC cDNA from the protonemal cells and analyzed the function as a bifunctional *ent*-kaurene synthase (HpDTC3).

Our current studies on identification and characteristics of cytochrome P450 monooxygenases responsible for the moss momilactones biosynthesis are in progress.

BIOCHEMICAL MECHANISM OF INDOLE-3-ACETIC ACID BIOSYNTHESIS IN *ARABIDOPSIS*

Kiyoshi Mashiguchi[1,2], Hitoshi Sakakibara[1] Yuji Kamiya[1] and Hiroyuki Kasahara[1]

[1]RIKEN Center for Sustainable Resource Science, 1-7-22, Tsurumi, Yokohama 230-0045, Kanagawa, Japan

[2]Corresponding author: kiyoshi.mashiguchi@riken.jp

Indole-3-acetic acid (IAA) is the naturally occurring auxin in plants. Understanding of the mechanism of IAA biosynthesis will allow us to manipulate its cellular levels in various plants. The indole-3-pyruvic acid (IPA) pathway, the main IAA biosynthesis pathway, has been established recently in *Arabidopsis.* However, the molecular mechanism of IAA biosynthesis has not yet been fully elucidated, especially for the biochemical function of YUCCA (YUC) flavin-containing monooxygenases that catalyze a rate-limiting step from IPA to IAA. In this study, we characterized six recombinant *Arabidopsis* YUC proteins from all major clades in the phylogenetic tree of YUCs. Moreover we reconstituted and analyzed the whole IPA pathway consisting of TRYPTOPHAN AMINOTRANSFERASE OF ARABIDOPSIS (TAA) and YUC *in vitro.* Our biochemical data implicate that the coordinated action of TAAs and YUCs may be important for the selective synthesis of IAA in plants.

GUMMOSIS IN BULBOUS PLANTS OF GRAPE HYACINTH, HYACINTH AND TULIP: FOCUS ON HORMONAL REGULATION AND CHEMICAL COMPOSITION OF GUMS

Kensuke Miyamoto[1,5], Toshihisa Kotake[2], Marian Saniewski[3], Junichi Ueda[4]

[1]Faculty of Liberal Arts and Sciences, Osaka Prefecture University, Japan

[2]Graduate School of Science and Engineering, Saitama University, Japan

[3]Research Institute of Horticulture, Poland

[4]Graduate School of Science, Osaka Prefecture University, Japan

[5]Corresponding author: miyamoto@las.osakafu-u.ac.jp

Gummosis, the process of the accumulation and exudation of gums mainly consisted of polysaccharides, has been known as a common response to various environmental stresses such as wounding, pathogen infections or insect attacks in some plant species. Hormonal regulation of gummosis has been intensively studied in stone-fruit species of the family Rosaceae, suggesting an important role of ethylene. We have found that jasmonic acid (JA) and methyl jasmonate (JA-Me) designated as jasmonates (JAs) induce gummosis as well as ethylene. In bulbous plants such as tulip, grape hyacinth and hyacinth, gummosis also takes place, however, little is known about the nature of gummosis. The purpose of this study is to clarify hormonal regulation of gummosis and chemical composition of gums from bulbous plants. In tulip shoots, JA-Me but not ethephon (2-chloroethylphosphonic acid) as ethylene-releasing compound induced gummosis, although both JA-Me and ethephon induced gummosis in bulbs. In grape hyacinth bulbs, ethephon induced gummosis, but JA-Me alone did not. JA-Me significantly enhanced ethephon-induced gummosis in grape hyacinth bulbs. In hyacinth bulbs, both JA-Me and ethephon induced gummosis, and these synergistically interacted on gummosis. Analyses of molecular mass distribution and sugar composition of gums suggested that gums of grape hyacinth and tulip were almost homogenous polysaccharides, being rich in Rha, Gal and Ara and uronic acids with an average Mw of 8.3 kDa and glucuronoarabinoxylans with an average Mw of ca. 700 kDa, respectively. Preliminary studies on the composition of hyacinth gums suggested that gums are pectic arabinogalactans with an average Mw of ca. 30 kD. These results suggests sugar metabolism producing gums is regulated by signal network of JAs and ethylene, especially by cross-signals between them, whereas sugar metabolisms relating to gummosis and key hormone to induce gummosis are different among species of bulbous plants.

INTRODUCTION

Gums are complexes of different substances but the most important constituents are polysaccharides (Boothby 1983). Induction and formation of gums designated as gummosis are found throughout the plant kingdom, especially in plants of the Rosaceae.

Since gummosis is induced by biotic and abiotic environmental factors such as bacterial and fungal infections, insect attacks, mechanical and chemical injuries, water stress, and other environmental stressors in some plant species, gums have been considered to play protective role in particular preventing entry of pathogens and insects to injured tissues and preventing moisture loss from damaged tissues (Olien and Bukovac 1982; Boothby 1983). Hormonal regulation of gummosis has been intensively studied in stone-fruit trees and fruits of the family Rosaceae (Olien and Bukovac 1982; Boothby 1983; Morrison et al. 1987). Since all these environmental factors mentioned above are considered to act via ethylene produced in plant tissues, and ethylene stimulates gummosis in stone-fruit species of the Rosaceae, ethylene is suggested to be a common factor involved in the induction of gummosis (Boothby 1983).

As well as ethylene, jasmonic acid (JA) and methyl jasmonate (JA-Me) designated as jasmonates (JAs) play an important role in signal transduction pathway in response to various stresses (Wasternack and Hause 2013). JAs also induce gummosis and interact with ethylene on gummosis in trees and fruits of various stone-fruit species of the family of Rosaceae such as plum, peach, cherry and apricot (Saniewski et al. 1998a, 2002, 2003; Ueda et al. 2003). In bulbous plants such as tulip (*Tulipa gesneriana*) and grape hyacinth (*Muscari armeniacum*), gummosis also takes place. In tulip bulbs, infection with *Fusarium oxysporum* f. sp. *tulipae* and the application of ethylene or JAs produced large quantities of gums, suggesting that pathogen-induced ethylene or JAs in plant tissues or these compounds produced by pathogens is involved in gummosis (see review, Saniewski et al. 2007). We have recently found that some hyacinth plants infected by *Fusarium oxysporum* showed the symptoms of gummosis in the plantations in Poland.

Beside tulips, the hormonal regulation of gummosis in bulbous plants has not been clear yet. In this paper we summarized hormonal regulation of gummosis and chemical composition of gums from bulbous plants based on our studies in tulips (Skrzypek et al. 2005a, b), grape hyacinth (Miyamoto et al. 2010, 2011) and hyacinth. Possible mode of action of these compounds in gummosis in bulbous plants will be also discussed.

MATERIALS AND METHODS

Plant materials and hormone treatment. The treatments of plant hormones were performed to bulbs of hyacinth (*Hyacinthus orientalis* L.). Hyacinth bulbs dug out from experimental field. After lifting, the bulbs were stored at room temperature (17~22 °C) until use. Intact bulbs were treated with lanolin (control), ethephon (2-chloroethylphosphonic acid, 2%, w/w), JA-Me (1.5%, w/w) and their mixture in lanolin paste (ca. 350 mg) on the basal plate of intact bulbs during the period of July to August. The treated bulbs were kept in room conditions at 17~22 °C. A lot of 10 to 12 bulbs were used in each treatment. After incubation, gummosis was observed optically, and photographed. Gums produced were collected, dried in natural room temperature and kept in the refrigerator until analysis. Since significant amount of gums was formed by simultaneous application of ethephon and JA-Me, gums formed by these phytohormones were subjected to sugar analysis.

Sugar analyses of gums in hyacinth bulb. Gums from hyacinth bulbs were dissolved in hot water, and then the solution was centrifuged at 3,000 g for 10 min. Almost all of gums were recovered in the supernatant. The contents of total sugars and uronic acids were determined by the phenol-sulfuric acid method (Dubois et al. 1956) using glucose (Glc) and carbazole-sulfuric acid method using glucuronic acid (GlcA) as standards (Galambos 1967), respectively. The molecular mass of gums was estimated by a gel permeation chromatography with a gel-permeation column (TSK-gel G5000PW, Tosoh Co. Ltd., Japan) according to the method of Wakabayashi et al. (1997). The weight-average molecular mass of gums was estimated according to the equation reported by Nishitani and Masuda (1981). Sugar composition of gums were analyzed by high performance anion-exchange chromatography with pulsed amperometric detection (HPAEC-PAD) using a Dionex DX-500 liquid chromatograph fitted with a CarboPac PA-1 column (4 x 250 mm, Dionex Japan, Osaka, Japan) as described by Ishikawa et al. (2000) and Konishi et al. (2008).

RESULTS AND DISCUSSION

Hormonal regulation of gummosis. Hormonal regulation of gummosis has been intensively studied, indicating an important role of ethylene (Boothby, 1983). As described in Introduction, the significance of the interaction of ethylene and JA-Me in gummosis has been reported (Saniewski et al. 1998a, 2002, 2003; Ueda et al. 2003).

In tulips (*Tulipa gesneriana* L.), both ethylene and JAs are capable to inducing gummosis in bulbs (Skrzypek et al. 2005a, b). On the contrary, exogenously applied ethephon, an ethylene-releasing compound, did not induce, but JA-Me (1%, w/w in lanolin) substantially induced gummosis in tulip shoots (Fig. 1A), suggesting that JAs are essential or principal factors to induce gummosis in tulips. Ethephon extremely stimulated gummosis induced by JA-Me in tulip shoots (Fig. 1B).

In grape hyacinth (*Muscari armeniacum* L.), the application of ethephon (1%, w/w in lanolin) to the basal plate of bulbs induced gummosis within several days after the application, gums being exuded around the basal plate, the place of the application, whereas the application of lanolin alone had no effect on gummosis (Miyamoto et al. 2010, 2011). Simultaneous application of JA-Me and ethephon extremely stimulated gummosis as compared with the case of ethephon alone (Fig. 2). Differently from the bulbs, neither the application of ethephon nor the simultaneous application of ethephon and JA-Me induced gummosis in inflorescence axes of grape hyacinth plants.

Hyacinth bulbs are injured by several bacterial and fungal diseases. We have recently found that some hyacinth plants infected by unknown pathogen showed the symptoms of gummosis in the plantations in Poland (Fig. 3B). The pathogen inducing gummosis was identified as *Fusarium oxysporum. Fusarium oxysporum* has been known to cause grey-brown lesions scattered over the bulb scales of hyacinth plants as well as typical basal rot of hyacinth. Since pathogen-induced ethylene or JAs in plant tissues or these compounds produced by pathogens is considered to be involved in gummosis (see review, Saniewski et al. 2007), it is possible that *Fusarium oxysporum* is capable to inducing gummosis via ethylene or JAs production in hyacinth bulbs. Both ethephon and JA-Me substantially induced gummosis in hyacinth bulbs. Simultaneous application of JA-

Me extremely stimulated ethephon-induced gummosis in hyacinth bulbs, and *vice versa*.

In tulip shoots, JA-Me was effective in gummosis but ethylene was not. Contrarily, in grape hyacinth bulbs ethephon was effective but JA-Me was not. In hyacinth bulbs, both ethephon and JA-Me were effective in gummosis. In these bulbous plants, extreme gummosis was induced by the simultaneous application of ethephon and JA-Me, suggesting that gummosis is regulated by signal network of JAs and ethylene, especially by cross-signals between them.

The mode of action of JAs to stimulated gum formation in bulbs in the presence of ethylene is not clear yet. JA-Me has been reported to stimulate 1-aminocyclopropane-1-carboxylic acid (ACC) synthase and ACC oxidase activities and/or ethylene production in various plant species (Emery and Reid 1996; Saniewski 1997). In tulip plants, JA-Me stimulated the evolution of ethylene and activity of ACC oxidase during gum induction (Saniewski and Wegrzynowicz-Lesiak 1994; 1995; Skrzypek et al. 2005a). In plum fruits JA-Me synergistically enhanced ethephon-induced gummosis, and *vice versa*. Simultaneous application of inhibitors of ethylene biosynthesis applied with JA-Me did not limit gummosis, but that of inhibitors of jasmonate biosynthesis together with ethephon significantly inhibited gummosis (Saniewski et al. 2004). Effectiveness of JA-Me to induce gummosis and to produce ethylene decreased as ripening stage of fruits. These suggest JAs together with endogenous levels of ethylene regulate gummosis. Studies on endogenous hormonal levels in response to the application of JA-Me or ethephon in bulbous plants will be required.

It is probable that JAs show synergistic effect in another way than only through stimulation of ethylene production in gummosis. Emery and Reid (1996) have reported that simultaneous application of ACC and JA-Me caused dramatic degradation of cell membrane as indicated by an increase in conductivity in sunflower seedlings. In tulip shoots, JA-Me substantially affected the sugar metabolism, resulting in the reduction of soluble sugars such as sucrose and reducing sugars (Skrzypek et al. 2005a). In oat coleoptile segments, JAs inhibited IAA-induced cell elongation by affecting the synthesis of cell wall polysaccharides (Ueda et al. 1994; 1995). JAs promoted abscission in bean petiole explants by affecting the metabolism of cell wall polysaccharides and cellulase activity (Ueda et al. 1996). Senescence stimulated by JAs in oat leaf segments was well correlated with changes in cell wall polysaccharides, especially cellulosic ones (Miyamoto et al. 2013). Modification of the sugar metabolism caused by JAs is possible to affect gummosis.

Studies on seasonal changes in gummosis in grape hyacinth bulbs revealed that ethephon-induced gummosis was not observed in April, but on the middle of May to June, gummosis in response to ethephon was found according to the advance of the season (Miyamoto et al. 2010). Then ethephon-induced gummosis reduced. Similar seasonal changes to induce gummosis were also observed in response to simultaneous application of JA-Me and ethephon. In plum fruits effectiveness of JA-Me and ethephon to induce gummosis decreased as ripening stage of fruits, gummosis-inducing activity of JA-Me being stronger than that of ethephon in mature fruits (Saniewski et al. 2004). JA-Me promoted ethylene production in green-colored immature plum fruits, but less in mature ones. These suggest that susceptibility to JAs or ethylene is dependent on the physiological conditions such as endogenous hormonal levels and sugar metabolism in plants as mentioned above.

Chemical composition of gums. Since gums are complexes of different substances but the most important constituents are polysaccharides, the molecular mass distribution of gums from tulip bulbs and grape hyacinth bulbs was analyzed by a gel permeation chromatography. Both gums from tulip and grape hyacinth were consisted of almost homogenous polysaccharides with an average molecular mass of ca. 700 kDa and 8.3 kDa, respectively (Fig. 4, Skrzypek et al. 2005a, Miyamoto et al. 2010).

Analysis of neural sugar composition of tulip gums revealed that majorities were Ara and Xyl, suggesting that tulip gums are mainly glucuronoarabinoxylans, whereas uronic acids has not been identified yet (Fig. 4A, Skrzypek et al. 2005a). On the other hand, in gums of grape hyacinth the molar ratio of Rha:Ara: Gal: GalA: GlcA was 25: 10: 40: 7: 15 (Fig. 4B). Grape hyacinth gums also contained proteins, the ratio of sugars and proteins being ca 94: 6. These suggest that grape hyacinth gums are mainly consisted of arabinogalactan proteins whose uronic acids are generally GlcA and its derivatives, but the possibility that the gums also contain rhamnogalacturonan, which is generally rich in Rha, Gal and GalA is not excluded (Miyamoto et al. 2010). Analyses of molecular mass and sugar composition of hyacinth gums indicated that majorities of sugars were Ara and Gal together with small amounts of Fuc, Rha and uronic acids, suggesting that hyacinth gums are pectic arabinogalactans with an average molecular mass of ca. 30 kDa (data not shown).

The chemical composition of the exuded gums demonstrated the gums of stone-fruit trees are complex branched hetero polysaccharides comprising residues of Gal, Ara and GlcA with other sugars also present in small or trace quantities (Boothby, 1983). On the other hand, analyses by a gel permeation chromatography revealed that gums from tulip, grape hyacinth and hyacinth were almost homogenous polysaccharides with an average molecular mass of ca. 700 kDa, 8.3 kDa and 30 kDa, respectively. These facts suggest that from the view point of molecular mass, sugar metabolism leading to gummosis in bulbous plants are simple as compared with those of stone-fruit species of Rosaceae whose molecular species of gums are quite different. However, sugar metabolisms leading gummosis are different between hyacinth, grape hyacinth and tulip based on the analysis of sugar composition.

In conclusion, JAs interact with ethylene in stimulation of sugar metabolism producing gums in bulbous plants, whereas sugar metabolisms and hormonal regulation relating to gummosis are different among species of bulbous plants, tulips, grape hyacinth and hyacinth.

Acknowledgements

The authors thank Prof. Takayuki Hoson and Drs. Kazuyuki Wakabayashi and Kouichi Soga (Osaka City University) for use the gel permeation chromatograph for the analysis of molecular mass distribution of gums and invaluable suggestions. This study was partially supported JSPA KAKENHI Grant number 24620008 (KM).

LITERATURE CITED

Boothby D. Gummosis of stone-fruit trees and their fruits. J Sci Food Agric 1983; 34: 1-7.

Dubois M, Gilles KA, Hamilton JK, Roberts PA, Smith F. Colorimetric method for determination of sugars and related substances. Anal Chem 1956; 28: 350-56.

Emery RJN, Reid DM. Methyl jasmonate effects on ethylene synthesis and organ-specific senescence in *Helianthus annuus* seedlings. Plant Growth Regul 1996; 18: 213-22.

Galambos JT. The reaction of carbazole with carbohydrates. I. Effect of borate and sulfamate on the carbazole color of sugars. Anal Biochem 1967; 19: 119-132.

Ishikawa M, Kuroyama H, Takeuchi Y, Tsumuraya Y. Characterization of pectin methyltransferase from soybean hypocotyls. Planta 2000; 210: 782-91.

Konishi T, Kotake T, Soraya D, Matsuoka K, Koyama T, Kaneko S, Igarashi K, Samejima M, Tsumuraya Y. Properties of family 79 β-glucuronidases that hydrolyze β-glucuronosyl and 4-*O*-methyl-β-glucronosyl residues of arabinogalactan-protein. Carbohydr Res 2008; 343: 1191-201.

Miyamoto K, Oka M, Uheda E, Ueda J. Changes in metabolism of cell wall polysaccharides in oat leaves during senescence: relevance to the senescence-promoting effect of methyl jasmonate. Acta Physiol Plant 2013; 35: 2675-83.

Miyamoto K, Kotake H, Sasamoto M, Saniewski M, Ueda J. Gummosis in grape hyacinth (*Muscari armeniacum*) bulbs: hormonal regulation and chemical composition gums. J Plant Res 2010; 123: 363-70.

Miyamoto K, Sasamoto M, Ueda J, Saniewski M. Interaction of ethylene and methyl jasmonate on gummosis in grape hyacinth (*Muscari armeniacum*) bulbs. Acta Hort 2011; 886: 341-47.

Morrison JC, Labavitch JM, Greve CL. The role of ethylene in initiating gum duct formation in almond fruit. J Amer Soc Hort Sci 1987; 112: 364-67.

Nishitani K, Masuda Y. Auxin-induced changes in the cell wall structure: Changes in the sugar composition, intrinsic viscosity and molecular weight distributions of matrix polysaccharides of the epicotyls cell wall of *Vigna angularis*. Physiol Plant 1981; 52: 482-94.

Olien W, Bukovac MJ. Ethephon-induced gummosis in sour cherry (*Prunus cerasus* L.) I. Effect on xylem function and shoot water status. Plant Physiol 1982; 70: 547-55.

Saniewski M. The role of jasmonates in ethylene biosynthesis. In: Kanellis AK et al. (eds) Biology and Biotechnology of the Plant Hormone Ethylene. Kluwer Academic Publishers, Dordrecht, 1997; pp 39-45.

Saniewski M, Wegrzynowicz-Lesiak E. Is ethylene responsible for gum formation induced by methyl jasmonate in tulip stem? J. Fruit Ornam Plant Res 1994; 2: 79-80.

Saniewski M, Wegrzynowicz-Lesiak E. The role of ethylene in methyl jasmonate-induced gum formation in stem of tulips. Acta Hortic 1995; 394: 305-13.

Saniewski M, Miyamoto K, Ueda J. Methyl jasmonate induces gums and stimulates anthocyanin accumulation in peach shoots. J Plant Growth Regul 1998a; 17: 121-124.

Saniewski M, Miyamoto K, Ueda J. Gum formation by methyl jasmonate in tulip shoots is stimulated by ethylene. J Plant Growth Regul 1998b; 17: 179-183.

Saniewski M, Miyamoto K, Ueda J. Gum induction by methyl jasmonate in fruits, stem and petioles of *Prunus domestica* L. Acta Hortic. 2004; 636: 151-58.

Saniewski M, Miyamoto K, Okubo H, Saniewska A, Puchalski J, Ueda J. Gummosis in tulip (*Tulipa gesneriana* L.): Focus on hormonal regulation and carbohydrate metabolism. Floriculture and Ornamental Biotechnology 2007; 1: 34-40.

Saniewski M, Ueda J, Miyamoto K, Urbanek H. Interactions between ethylene and other plant hormones in regulation of plant growth and development in natural conditions and under abiotic and biotic stresses. In: Vendrell M et al. (eds) Biology and Biotechnology of the Plant Hormone Ethylene III, IOS Press (ISBN 1 58603 346 8); 2003; pp 263-70.

Saniewski M, Ueda J, Horbowicz M, Miyamoto K, Puchalski J. Gum in apricot (*Prunus armeniaca* L.) shoots induced by methyl jasmonate. Acta Agrobotanica 2002; 54: 27-34.

Saniewski M, Ueda J, Miyamoto K, Horbowicz M, Puchalski J. Hormonal control of gummosis in *Rosaceae.* J Fruit and Ornamental Plant Research 2006; 14(Suppl. 1): 139-43.

Skrzypek E, Miyamoto K, Saniewski M, Ueda J. Jasmonates are essential factors inducing gummosis in tulips: mode of action of jasmonates focusing on sugar metabolism. J Plant Physiol 2005a; 162: 495-505.

Skrzypek E, Miyamoto K, Saniewski M, Ueda J. Identification of jasmonic acid and its methyl ester as gum-inducing factors in tulips. J Plant Res 2005b; 118: 27-30.

Ueda J, Miyamoto K, Aoki M. Jasmonic acid inhibits the IAA-induced elongation of oat coleoptile segments: a possible mechanism involving the metabolism of cell wall polysaccharides. Plant Cell Physiol 1994; 35: 1065-70.

Ueda J, Miyamoto K, Kamisaka S. Inhibition of the synthesis of cell wall polysaccharides in oat coleoptile segments by jasmonic acid: Relevance to its growth inhibition. J Plant Growth Regul 1995; 14: 69-76.

Ueda J, Miyamoto K, Hashimoto M. Jasmonates promote abscission in bean petiole explants: Its relationship to the metabolism of cell wall polysaccharides and cellulase activity. J Plant Growth Regul 1996; 15: 189-95.

Ueda J, Miyamoto K, Saniewski M. Gum formation and leaf abscission in ornamental Japanese cherry (*Prunus yedoensis*): A possible role of ethylene and jasmonates in these processes. In: Vendrell M et al. (eds) Biology and Biotechnology of the Plant Hormone Ethylene III, IOS Press (ISBN 1 58603 346 8), 2003; pp 263-70.

Wakabayashi K, Hoson T, Kamisaka S. Suppression of cell wall stiffening along coleoptiles of wheat (*Triticum aesativum L.*) seedlings grown under osmotic stress conditions. J Plant Res 1997; 110: 311-16.

Wasternack C, Hause B. Jasmonates: Biosynthesis, perception, signal transduction and action in plant stress response, growth and development. An update to the 2007 review in Annals of Botany. Annal Bot 2013; 111: 1021-1058.

A.

	Gummosis in tulip	
Treatments	Bulbs	Shoots
None	–	–
JA-Me	+	++
Ethephon	+	–
JA-Me + ethephon	++++	++++

B.

Figure 1. Effects of ethephon and/or JA-Me on gummosis in tulip plants. A. Formation of gums was observed optically 7 days after treatment of JA-Me (1%, w/w in lanolin) and/or ethephon (1%, w/w in lanolin). Gum formation was observed 7 days after treatment: – = no gum; + to ++++ = increasing gum formation. Results were of Skrzypek et al. (2005a, b) with some modifications. B. Gummosis in tulip shoots induced by the simultaneous application of JA-Me and ethephon.

A.

	Gummosis in grape hyacinth	
Treatments	Bulbs	Inflorescence axes
None	–	–
JA-Me	–	–
Ethephon	+	–
JA-Me + ethephon	++++	–

B.

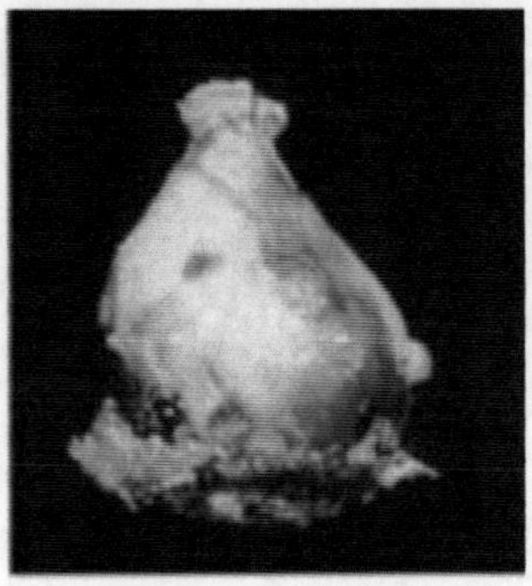

Figure 2. Effects of ethephon and/or JA-Me on gummosis in grape hyacinth plants. A. Formation of gums was observed optically 7 days after treatment: – = no gum; + to ++++ = increasing gum formation. Results were of Miyamoto et al. (2010, 2011) with some modifications. B. Gummosis in grape hyacinth bulbs induced by the simultaneous application of JA-Me and ethephon.

A.

Treatments	Gummosis in hyacinth	
	Bulbs	Inflorescence axes
None	–	NE
JA-Me	–	NE
Ethephon	+	NE
JA-Me + ethephon	++++	NE

B.

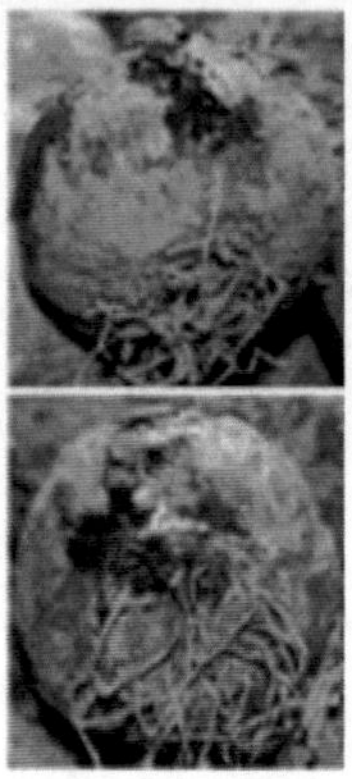

Figure 3. Effects of ethephon and/or JA-Me on gummosis in hyacinth plants. A. JA-Me (1.5%, w/w in lanolin) and ethephon (2%, w/w in lanolin) were used. Formation of gums was observed optically: – = no gum; + to ++++ = increasing gum formation; NE = not examined. B. Gummosis induced by the injection with *Fusarium oxysporum* in plantations.

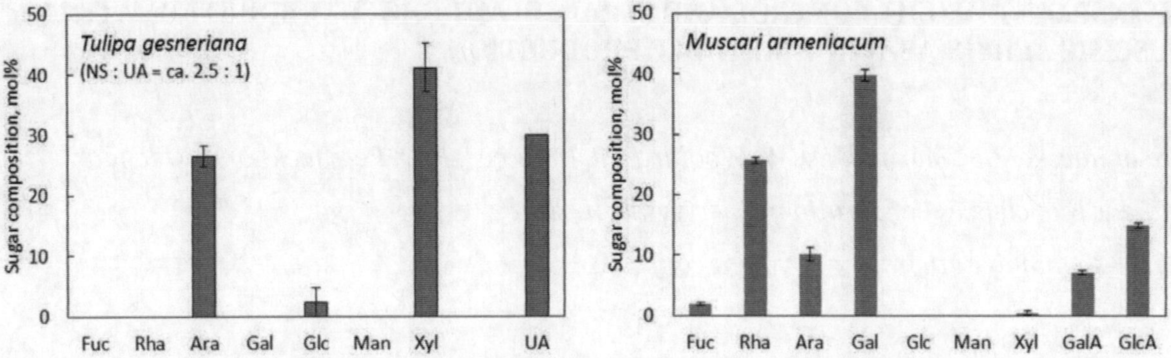

Figure 4. Sugar composition of gums from tulip and grape hyacinth bulbs treated with JA-Me and ethephon. A: After hydrolysis of tulip gums with 2N trifluoroacetic acid, soluble neutral sugars together with inositol as an internal standard were reduced with sodium borohydride and acetylated with acetic anhydride. Qualitative and quantitative analyses of alditol acetates were carried out using a gas-liquid chromatography fitted with a hydrogen flame ion detector according to the method reported previously (Skrzypek et al. 2005a). B: Sugar composition of grape hyacinth gums hydrolyzed was analyzed by HPAEC-PAD using a Dionex DX-500 liquid chromatograph fitted with a CarboPac PA-1 column (4 x 250 mm, Dionex Japan, Japan) as described by Ishikawa et al. (2000) and Konishi et al. (2008).

TRANSPLANT HEIGHT CONTROL AND "TRANSPLANT SHOCK" REDUCTION WITH S-ABSCISIC ACID (S-ABA) IN VEGETABLE PRODUCTION

Jozsef Racsko[1,2], F. Marmor[1], D. W. Woolard[1], R. Hopkins[1], P. D. Petracek[1] and J. Pienaar[1]

[1]Valent BioSciences Corporation, Libertyville, IL, USA

[2]Corresponding author: jracs@valent.com

Transplanting is a standard cultural practice in vegetable production to improve seedling survival and cropping characteristics (earliness, yield, crop quality). The major objective of vegetable seedling production for transplanting is to produce a plant that has a compact shoot and well-developed, strong root system that provides a better chance of survival when it is moved from the protected environment to the field. S-ABA has been proven to successfully reduce undesirable excess shoot growth in the greenhouse with an increase in root-to-shoot ratio and improve seedling hardiness in a wide range of species. In tomato seedlings, foliar application of 750-1,000 ppm ABA showed significant growth reduction without visible phytotoxic effects. In pepper, 250-500 ppm appeared to be the optimum concentration. Seedlings taken from the greenhouse and planted in the field often suffer transient water stress (i.e., transplant shock) due to root injury during transplanting and disturbed root-soil contact primarily in exposure to high evapotranspiration demand. Recently, foliar applications of S-ABA have gained interest in the vegetable industry as a method to improve post-transplant stress tolerance and increase transplant stand establishment. In field trials, foliar applications of 150-1,000 ppm S-ABA significantly improved post-transplant survival rate, increased fruit size and yield in tomato.

THE COMMERCIAL DEVELOPMENT OF AMINOETHOXYVINVYLGLYCINE (AVG) – A HISTORICAL PERSPECTIVE

Warren Shafer[1,2]

[1]*Valent Biosciences Corp, Libertyville, IL, USA*

[2]*Corresponding author: warren.shafer@valentbiosciences.com*

Dr. Shafer will deliver the PGRSA 2014 Invited Historical Perspectives Presentation. He will review the story behind AVG and how it was ultimately registered for use on apples in 1997. While Abbott Labs did not discover AVG, they did made it a commercial reality for crop production. The product development journey involved many talented and dedicated people who persevered through significant challenges and found a way to make good things happen. And the story of AVG continues to unfold 17 years later with new uses continuing to be developed. Dr. Shafer played a major role in the commercialization of AVG and his presentation will focus on the Discovery, History of Scientific Research and Commercial Development of AVG.

DETERMINATION OF ROOT GROWTH IN FIELD-GROWN AND PGR-TREATED WINTER WHEAT

Wilhelm Rademacher[1,2,3]

[1]*BASF Global Research Crop Protection, 67117 Limburgerhof, Germany*

[2]*Corresponding author: radifax@t-online.de*

[3]*Current address: Austrasse 1, 67117 Limburgerhof, Germany*

INTRODUCTION

The plant growth regulator Medax® Top (300 g/l mepiquat chloride + 50 g/l prohexadione-Ca) has been introduced in different countries as an anti-lodging product in cereals since 2005 (trade name in the UK and in Ireland: Canopy®. Whereas morphological effects on shoot parameters (e.g. shoot length and stem diameter) caused by this product could be determined without major difficulties, obtaining relevant data on root growth under field conditions has been a challenge. The root system of a cereal or other plant species is of paramount importance not only because of absorbing water and nutrients. It also represents the plant's "foundation", which helps to keep the shoot upright and which is, therefore, highly important to reduce the risk of lodging in intense cereal production. Furthermore, roots serve as "bioreactors" and produce cytokinins and other important compounds for export into the shoot.

In principle, inhibitors of GA biosynthesis, such as mepiquat chloride and prohexadione-Ca, enable a plant to shift assimilates no longer used for shoot growth into the root system. The roots might then reach deeper soil layers and might form a better "root plate". Uptake of water and nutrients would be improved under conditions of drought. Stems would also be kept upright better in a soft soil, soaked with water after intense rainfall.

MATERIALS AND METHODS

An area with a uniform light soil (loamy sand) was selected for growing winter wheat (cv. 'Bussard') for the analysis of root growth and other parameters. The previous crop had been radishes, which minimized the presence of potentially interfering plant remnants in the soil. Plants were grown in plots of, each, 9 m^2 with six replications in a randomized block design. Two plots were used to conduct root analyses. The other plots were used to determine shoot height, incidence of lodging, and seed yield. The following variants were chosen as treatments:

a) Untreated control.

b) 1,000 ml of Medax® Top + 500 g of ammonium sulfate per hectare applied at growth stage 31 BBCH (first node at least one 1 cm above tillering node).

c) Two times 750 ml Medax® Top + 500 g of ammonium sulfate per hectare applied at growth stages 31 and 37 (flag leaf just visible) BBCH.

Root analyses were conducted at anthesis, when root growth is known to terminate. At this time, the soil moisture content was at approximately 43% of field capacity. Typical plants with three to four ear-bearing stems were selected in each experimental variant. Two different approaches were chosen for root determination:

a) Roots with the surrounding soil were carefully dug out to a depth of approximately 40 cm as one block. After soaking in water, soil particles were gently washed off from the roots. Finally, aliquots of the respective root samples were scanned to determine total length and other root parameters by using WinRhizo® software (Regent Instruments Inc., Quebec, Canada). (10 repeats per variant.)

b) A 20 kg spring balance (PESOLA AG, Baar, Switzerland) was attached to the basal parts of a wheat plant. The plant was then pulled out of the soil, and the required force read from the drag pointer of the spring balance (1 kg = 1 kp = 9.8 N). (20 repeats per variant.)

Results and Discussion

As expected, the first method was very labor-intensive and required a high degree of diligence. Two persons had to work approximately five days to analyze 30 samples. In contrast, 60 measurements with the root-pulling resistance method could be performed by one person in less than one day. The obtained data (Table 1) demonstrate that especially the split application of Medax® Top reduces shoot length in winter wheat and has a significant effect on lodging, which results in increases in seed yield by approximately 13%. The split treatment resulted also in increased root growth. This could be demonstrated both by the highly laborious measurement of root length by scanning and by the "quick and easy" determination of root pulling resistance. Because of its simplicity, the latter method might be of interest to obtain root data for orientation in a number of cases.

Table 1. Effects of Medax Top on shoot and root growth, lodging incidence, seed yield, and root-pulling resistance in field-grown winter wheat, cv. Bussard.

Treatment	Shoot length (cm)	Lodging (% of area)	Seed yield (t/ha)	Root length (m/ear-bearing stem)	Root-pulling resistance (N/ear-bearing stem)
Control	120a	82a	6.65a	5.27ab	22.6a
Medax Top 1,000 mL/ha	114ab	65ab	6.77ab	5.05a	25.5ab
Medax Top 2x 750 mL/ha	105b	0c	7.54c	6.58ab	35.3c

RESPONSE OF *Amphicarpaea bracteata* (*L.*), THE DAKOTA PEA, TO PLANT GROWTH REGULATORS AND PHOTOPERIOD

Emily Bethke[1] *and Sonja L. Maki*[1,2]

[1]*Department of Plant and Earth Science, University of Wisconsin-River Falls, 310 N. 3rd Street, River Falls, WI 54022*

[2]*Corresponding author: sonja.maki@uwrf.edu*

Amphicarpaea bracteata (L.) is an edible annual woodland legume native to Midwestern and Eastern parts of North America. The plant inhabits woods and thickets and is known by several common names including Dakota Pea, Hog Peanut, and Pea Vine (Kindscher, 1987). This close relative of the soybean has been used as a staple in Native American diets and, in areas where the Dakota Pea grows wild, field mice rely heavily on the underground seeds for food (Kindscher, 1987). This plant has been under research as potential cover crop, due to its nitrogen fixing nature (Kindscher, 1987).

Amphicarpaea comes from the Greek language meaning "with two kinds of fruit" referring to the fruits the plant produces both above and below ground. The plant is capable of producing several distinct classes of seed sizes that aid in the species survival (Kindscher, 1987). Three different inflorescence morphologies are produced on the plant. In this study, the plant response to growth regulators was assessed under both long- and short-day lengths.

The objectives of this study were to investigate the effects of photoperiod on plant growth and development and to determine the effects of a gibberellin biosynthesis inhibitor (Prohexadione-Ca) on plant growth. Aerial seeds were collected from plants found in the wild in western Wisconsin. The seeds were sterilized in 15% bleach, followed by two three minute rinses in distilled water. Seeds were imbibed in 100 or 200 ppm Prohexadione-Ca (a gibberellin biosynthesis inhibitor; Rademacher (2000)) prior to a 3 day cold treatment (4C). Seeds were sown in a peat-based media and grown in a greenhouse. Long days (16h) were provided with with supplemental HID lighting and short days were provided by an automated short day cloth greenhouse system (9h). Growth data was analyzed with JMP11 (SAS)

Seeds treated with 100 ppm Prohexadione-Ca produced plants less than half the approximately one-half the normal height growth when compared to the control group (Fig. 2). Seeds treated with 200 ppm Prohexadione-Ca produced plants with even less growth. When *Amphicarpaea bracteata (L.)* is grown under natural long days in the Midwest, it flowers in late July. Prohexadione-Ca treatment affected plant height similarly in short days (data not shown), however plants flowered early, at the first node, and only one-seeded pods were produced, in contrast to the 3-4 seeded pods produced in upper regions of long day grown plants.

These one-seeded pods were not observed on long-daygrown plants, however more research on plants growing in their native environment is necessary to determine whether this fourth type of flowering and fruiting pattern occurs in nature.

The results presented in this research provide some basic plant growth regulator responses of a wild legume which can be used for teaching phenotypic plasticity of wild crop relatives in a laboratory setting. The results from the gibberellin synthesis inhibitor studies could also be of interest to plant breeders interested in the domestication syndrome. Future studies will utilize this data to assist in determining the genetic regulation of seed size as it relates to inflorescence type and the regulation of flowering.

Literature Cited

Kindscher, Kelly. Edible Wild Plants of the Prairie: An Ethnobotanical Guide. Lawrence, Kan.: U of Kansas, 1987.

W., Rademacher. "Growth Retardants: Effects on Gibberellin Biosynthesis and Other Metabolic Pathways." Annual Rev Plant Physiol Plant Molecular Biology 51 (2000): 501-31.

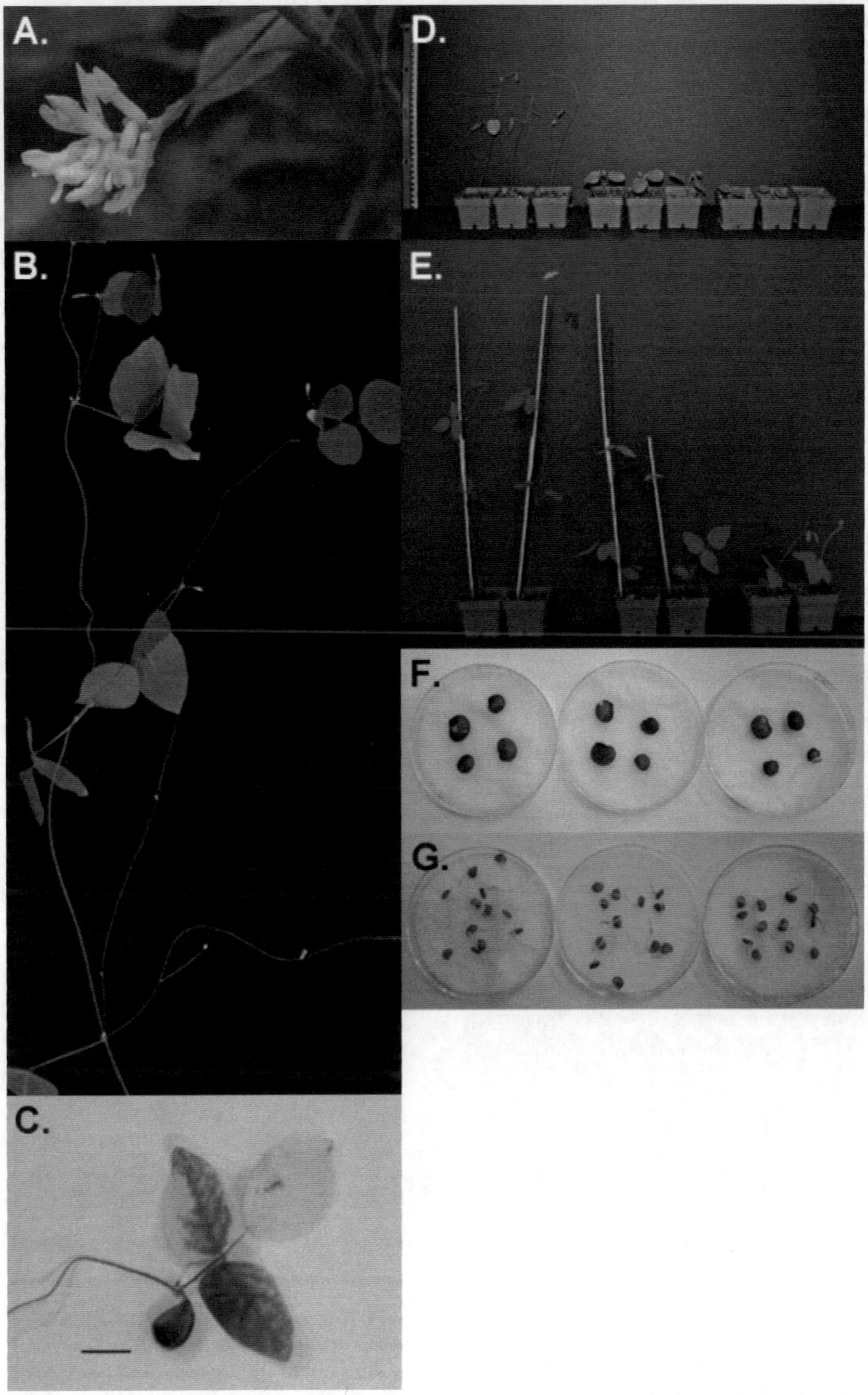

Figure 1. Photographs of *Amphicarpaea bracteata* (L.). A. Aerial inflorescence from a (B.) whole plant (C) one-seeded pod produced under short days (D, E) plants obtained from seeds treated with 0, 100 or 200 ppm Prohexadione development, (F) large underground seeds (G) small aerial seeds (treated with 0, 100 and 200 ppm Prohexadione-Ca). Note the vine like growth habit of the plant, and the long axillary branches. Blue bar = 10 cm; black bar = 1 cm.

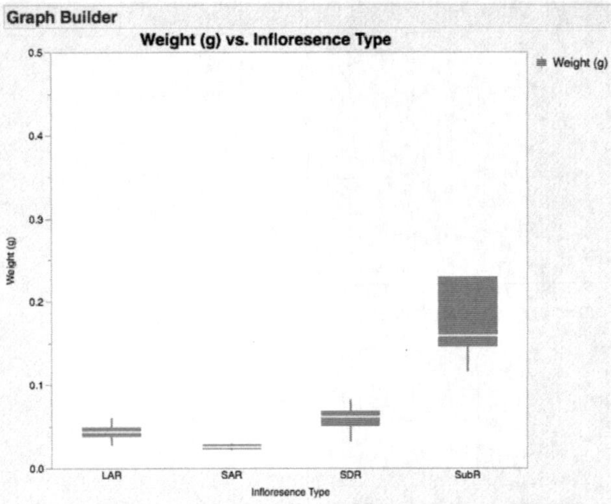

Figure 2. Box plot and statistical analysis of seed dry weights (g) weights from seeds obtained from fruits from Long Aerial Rcemes (LAR), Short Aerial Racemes (SAR), Short Day Racemes, and Subterranean Racemes (SubR).

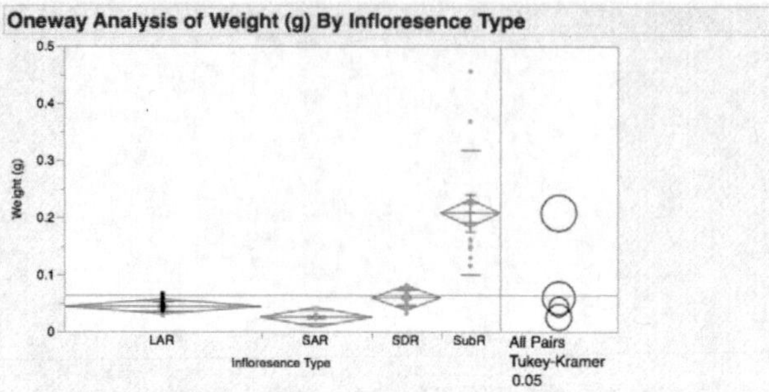

Oneway Anova

Summary of Fit

Rsquare	0.689712
Adj Rsquare	0.677778
Root Mean Square Error	0.039772
Mean of Response	0.063601
Observations (or Sum Wgts)	82

Analysis of Variance

Source	DF	Sum of Squares	Mean Square	F Ratio	Prob > F
Infloresence Type	3	0.27425226	0.091417	57.7931	<.0001*
Error	78	0.12338089	0.001582		
C. Total	81	0.39763315			

Means for Oneway Anova

Level	Number	Mean	Std Error	Lower 95%	Upper 95%
LAR	37	0.043962	0.00654	0.03095	0.05698
SAR	21	0.025233	0.00868	0.00795	0.04251
SDR	13	0.059431	0.01103	0.03747	0.08139
SubR	11	0.207836	0.01199	0.18396	0.23171

Std Error uses a pooled estimate of error variance

Means and Std Deviations

Level	Number	Mean	Std Dev	Std Err Mean	Lower 95%	Upper 95%
LAR	37	0.043962	0.009024	0.00148	0.04095	0.04697
SAR	21	0.025233	0.002802	0.00061	0.02396	0.02651
SDR	13	0.059431	0.014527	0.00403	0.05065	0.06821
SubR	11	0.207836	0.108517	0.03272	0.13493	0.28074

Means Comparisons

Comparisons for all pairs using Tukey-Kramer HSD

Confidence Quantile

q*	Alpha
2.62529	0.05

LSD Threshold Matrix

Abs(Dif)-HSD

	SubR	SDR	LAR	SAR
SubR	-0.04452	0.10563	0.12802	0.14374
SDR	0.10563	-0.04095	-0.01820	-0.00265
LAR	0.12802	-0.01820	-0.02428	-0.00980
SAR	0.14374	-0.00265	-0.00980	-0.03222

Positive values show pairs of means that are significantly different.

Connecting Letters Report

Level			Mean
SubR	A		0.20783636
SDR		B	0.05943077
LAR		B	0.04396216
SAR		B	0.02523333

Levels not connected by same letter are significantly different.

Ordered Differences Report

Level	- Level	Difference	Std Err Dif	Lower CL	Upper CL	p-Value
SubR	SAR	0.1826030	0.0148029	0.143741	0.2214649	<.0001*
SubR	LAR	0.1638742	0.0136584	0.128017	0.1997315	<.0001*

STUDIES ON THE EFFICACY OF A NEW FORMULATION OF UNICONAZOLE-P FOR USE IN CALIFORNIA AVOCADO PRODUCTION

Carol J. Lovatt[1] and Timothy M. Spann[2,3]

[1]University of California, Riverside, CA, USA

[2]California Avocado Commission, Irvine, CA, USA

[3]Corresponding author: tspann@avocado.org

Uniconazole-P is used in 'Hass' avocado production to stop vegetative shoot growth at the apex of indeterminate floral shoots to increase fruit set and yield and after pruning to maintain tree size, especially in high-density plantings. Uniconazole-P has the potential to reduce pruning costs, but also to reduce fruit size and increase fruit drop. Depending on crop load, reducing vegetative shoot growth in spring or summer could mitigate or initiate alternate bearing. The objectives were to determine the effectiveness of a new formulation of Uniconazole-P developed for use in California to stop shoot growth (i.e., what proportion of shoots stop growing and for how long) when applied in spring, summer, fall or winter and to determine its effects on yield, including fruit size. Consistent with its role as a GA biosynthesis inhibitor, summer-applied Uniconazole-P had a greater effect on internode elongation (~4 weeks) than on production of new nodes by the apical meristem (~2 weeks). At the end of 8 weeks, shoot length was reduced by up to 4 cm, compared to untreated controls, but results were inconsistent. Fall-applied Uniconazole-P produced similar results at the end of 8 weeks in both orchards reducing shoot growth by 2 cm or less. Winter treatments significantly reduced shoot growth for all shoots from week 8 through 12. Net reduction in length was ≤ 2 nodes and ≤ 4 cm relative to the untreated control. In contrast to the summer and fall applications, the winter application had a greater effect on shoot apical meristem growth than internode elongation.

EFFECTS OF ELEVATED AMBIENT PRESSURE ON THE RATES OF NET PHOTOSYNTHESIS AND DARK RESPIRATION

Motoki Yonekura[1,3], Hiroyuki Takeishi[1], Hirotaka Awata[1], Jun Hayashi[1], Akio Kobayashi[1], Yoshihiro Koshino-Kimura[1], Takashi Machimura[1], Fumiteru Akamatsu[1] and Atsushi Okazawa[2]

[1]Osaka University, Osaka, Japan

[2]Osaka Prefecture University, Osaka, Japan

[3]Corresponding author: yonekura@combu.mech.eng.osaka-u.ac.jp

In this study, the effect of environmental stress on the net photosynthesis was investigated under artificially controlled environment. In particular, an effect of elevated ambient pressure on the CO_2 exchange rate in photosynthesis of *Arabidopsis thaliana*, grown in an aseptic culture condition was discussed. To evaluate the overlapping effect of Arabidopsis leaves in a petri dish on the net photosynthesis and on the dark respiration, plants were grown in the different shoot density (20 and 40 shoot per 90 mm diameter dish). In this experiment the shoot density was used instead of leaf area index (LAI). The pressure container, made of acryl and aluminum, was used for the experiments. Using fluorescent lights as a light source, PPFD was controlled in the range of 0 (dark period) - 200 mmol m^{-2} s^{-1} (light period). Temperature in the pressure container was kept at 24°C in all experiments. The vessel inner pressure such as 0.1 (CO2 partial pressure: ca. 40 pa), 0.2 (CPP: ca.80) and 0.3 (CPP: ca. 120) MPa was offered using three different gas cylinders with different CO2 content. A non dispersive infra-red gas analyzer was used to measure CO_2 exchange rate. The result showed that the rate of net photosynthesis as well as the rate of dark respiration per the unit leaf area did not change, while the number of shoots in the petri dish was different. This result indicated that the shoot density has only little effect on the rates of net photosynthesis and dark respiration under elevated ambient pressures. In addition, the rate of net photosynthesis increased significantly, and the rate of dark respiration did not change under the elevated CO_2 partial pressure. On the contrary, when the CO_2 partial pressure was kept constant (ca.40 Pa) in the different increased pressures, the rate of net photosynthesis decreased and the rate of dark respiration increased. This may suggest that elevated ambient pressure has a stress to the growth of plants. It is accepted that an increase of the rate of net photosynthesis influences on the growth of plants, and the fact that the net photosynthesis rate increased under the elevated pressure suggested that elevated ambient pressure has a positive effect on the growth of plants.

Through this experiment, we have found various interesting physiological phenomena in the different plant species grown under such a unique condition.

OPTIMIZATION OF *ZINNIA MARYLANDICA* PRODUCTION IN THE GREENHOUSE USING LED LIGHTING WITH VARIABLE LIGHT RATIOS

James P. Byrtus[1], Kalli M. Egan[1] and M. Melanie Yelton[1,2]

[1]LumiGrow, 8 Digital Drive, Suite 200, Novato, CA 94949 USA

[2]Corresponding author: melanie@lumigrow.com

There has been resurgence in investigation of the how different intensities and ratios of light effects plant growth. Using *Zinnia marylandica* as a model organism and LED lighting with independently controlled red, blue, and white light we evaluate four different light treatments in the greenhouse. In each condition natural light was supplemented with 250 µmol $m^{-2}s^{-1}$ of light. White light was provided at a constant level of 20 µmol $m^{-2}s^{-1}$ blue light was provided at 0, 20, 40 and 60 µmol $m^{-2}s^{-1}$, and red light was added to bring light to the final level of 250 µmol $m^{-2}s^{-1}$. Three additional treatments were evaluated: no supplemental lighting (NSL), a LumiGrow Pro 325 fixture at full power (FP), and a 400 watt HPS. Results demonstrated that addition of increasing levels of blue light yielded a shorter plant, with a similar numbers of flower structures when compared to control plants grown under HPS. Results indicate that in the greenhouse environment supplemental light regimes can be created to control plant growth at different stages.

The need for Utilizing PGRs in Ornamental Cropping Systems

Dave Barcel[1,2]

[1]*OHP Inc., Genesee, WI, USA*

[2]*Corresponding author:* dbarcel@ohp.com

Ornamental cropping systems have changed dramatically over the past 50 years. The 1960s was the beginning of several advances in new varieties of ornamentals, soilless media mixes, double poly house as well as Horticultural chemical advancements such as plant growth regulators (PGRs) Several crops such as geranium, chrysanthemum, hydrangea and poinsettia were very vigorous in growth and would consume much bench or floor space if the growth was not held in check. Those earlier days of production utilized pinching to enhance crop quality by inducing more lateral breaks (hydrangea, mums and poinsettia) however, pinching also served to restrain un-wanted growth as well.

The advent of PGRs in the late 60s and early 70s brought about a new set of tools for growers to use in plant growth management. Products such as Cycocel (Chlormequat chloride) and B-Nine (Daminozide) provided significant growth reduction by inhibiting gibberellin synthesis, most importantly Ga3. In the coming years of the 80s and 90s other PGR's of significance came to market such as ARest (Ancymidol) Bonzi (Paclobutrazol) Sumagic (Uniconazole) and Topflor (Fluprimidol). These newer triazole based chemicals offered extremely active plant response in growth reduction, which allowed for use on bulbs, perennials and woody ornamentals, crops known for difficult to control growth. By the 2000s growers wanted plants that would develop better branching habits, which produced a fuller plant and offered more blooms per plant. This need resulted in a new set of PGR's known for inducing lateral bud break or new bud production. Products such as Florel (Ethephon) Configure (6-Ba) and Augeo (Dikegulac) triggering plant responses that increased branching which resulted in fuller plants for the consumer.

INDOLE-3-ACETIC ACID AND PHENYLACETIC ACID ARE TWO NATURAL AUXINS IN PLANTS WITH DISTINCT TRANSPORT CHARACTERISTICS

Hiroyuki Kasahara[1,2]

[1]*RIKEN Center for Sustainable Resource Science, Yokohama, Japan*

[2]*Corresponding author: kasahara@riken.jp*

The phytohormone auxin plays a central role in many aspects of plant growth and development, including embryogenesis, organ formation, and tropism (1). Auxin regulates cell elongation and differentiation in a concentration-dependent manner. A naturally occurring auxin, indole-3-acetic acid (IAA), is mainly synthesized from Trp *via* indole-3-pyruvate (IPA) by TRYPTOPHAN AMINOTRANSFERASE OF ARABIDOPSIS (TAA) and YUCCA (YUC) flavin-containing monooxygenases in *Arabidopsis* (2-6). The *TAA* and *YUC* genes probably have crucial roles in the spatiotemporal regulation of auxin concentration in plants (4,5,7-9). Recent studies demonstrate that L-kynurenine and yucasin are specific inhibitors for auxin biosynthesis in plants; L-kynurenine strongly inhibits the production of IPA by TAA and yucasin inhibits the conversion of IPA to IAA by YUC, respectively (10,11). These studies show that TAA and YUC enzymes are important targets in the development of herbicides and plant growth regulators.

It has been demonstrated that various conjugation and degradation enzymes also play important roles in the regulation of IAA concentration (12). Several *GRETCHEN HAGEN 3* (*GH3*) genes encode enzymes that catalyze the formation of IAA-amino acid conjugates (13). IAA CARBOXYL METHYLTRANSFERASE 1 (IAMT1) catalyzes the methylation of the carboxyl group, which inactivates IAA to its methyl ester (MeIAA) (14). UGT84B1 inactivates IAA to its glucoside (IAA-Glc) (15). UGT74E2 catalyzes the glucosylation of indole-3-butylic acid (IBA), which is probably synthesized from IAA in plants (16). Moreover, recent study demonstrates that DIOXYGENASE FOR AUXIN OXIDATION (DAO), a 2-oxoglutarate-dependent dioxygenase, catalyzes the conversion of IAA to 2-oxindole-3-acetic acid (OxIAA) (17). OxIAA has the weak auxin activity (18) and is further inactivated to its glucoside (OxIAA-Glc) by UGT74D1 in *Arabidopsis* (19). The *DAO* gene is essential for anther dehiscence, pollen fertility, and seed initiation in rice (17), thus, the OxIAA pathway would be a good target for the development of agrochemicals.

Phenylacetic acid (PAA) has been recognized as another natural auxin for more than 40 years, but its role in plant growth and development remains unclear. I will show that PAA may play an important role in the regulation of plant growth and development. PAA is widely distributed in both vascular and nonvascular plants, and the endogenous concentrations of PAA are higher than those of IAA in various plant species. The *YUC* family probably functions in PAA biosynthesis and the *GH3* family inactivates PAA through conjugation with amino acids. An important difference between IAA and PAA is that PAA is not transported actively and directionally. IAA and PAA can regulate the same set of auxin responsive genes but also various genes independently, suggesting that these auxins have overlapping but distinct regulatory roles in plants.

LITERATURE CITED

Davies, P. J. (2004) The plant hormones: their nature, occurrence, and functions. in *Plant Hormones - biosynthesis, signal transduction, action!* (Davies, P. J. ed.), Kluwer Academic Publishers, Dordrecht. pp 1-15

Mashiguchi, K., Tanaka, K., Sakai, T., Sugawara, S., Kawaide, H., Natsume, M., Hanada, A., Yaeno, T., Shirasu, K., Yao, H., McSteen, P., Zhao, Y., Hayashi, K., Kamiya, Y., and Kasahara, H. (2011) The main auxin biosynthesis pathway in *Arabidopsis. Proc Natl Acad Sci USA* 108, 18512-18517

Won, C., Shen, X., Mashiguchi, K., Zheng, Z., Dai, X., Cheng, Y., Kasahara, H., Kamiya, Y., Chory, J., and Zhao, Y. (2011) Conversion of tryptophan to indole-3-acetic acid by TRYPTOPHAN AMINOTRANSFERASES OF *ARABIDOPSIS* and YUCCAs in *Arabidopsis. Proc Natl Acad Sci USA* 108, 18518-18523

Tao, Y., Ferrer, J. L., Ljung, K., Pojer, F., Hong, F., Long, J. A., Li, L., Moreno, J. E., Bowman, M. E., Ivans, L. J., Cheng, Y., Lim, J., Zhao, Y., Ballare, C. L., Sandberg, G., Noel, J. P., and Chory, J. (2008) Rapid synthesis of auxin via a new tryptophan-dependent pathway is required for shade avoidance in plants. *Cell* 133, 164-176

Stepanova, A. N., Robertson-Hoyt, J., Yun, J., Benavente, L. M., Xie, D. Y., Doležal, K., Schlereth, A., Jurgens, G., and Alonso, J. M. (2008) *TAA1*-mediated auxin biosynthesis is essential for hormone crosstalk and plant development. *Cell* 133, 177-191

Stepanova, A. N., Yun, J., Robles, L. M., Novak, O., He, W., Guo, H., Ljung, K., and Alonso, J. M. (2011) The *Arabidopsis* YUCCA1 flavin monooxygenase functions in the indole-3-pyruvic acid branch of auxin biosynthesis. *Plant Cell* 23, 3961-3973

Cheng, Y., Dai, X., and Zhao, Y. (2006) Auxin biosynthesis by the YUCCA flavin monooxygenases controls the formation of floral organs and vascular tissues in *Arabidopsis. Genes Dev* 20, 1790-1799

Cheng, Y., Dai, X., and Zhao, Y. (2007) Auxin synthesized by the YUCCA flavin monooxygenases is essential for embryogenesis and leaf formation in *Arabidopsis. Plant Cell* 19, 2430-2439

Yamada, M., Greenham, K., Prigge, M. J., Jensen, P. J., and Estelle, M. (2009) The TRANSPORT INHIBITOR RESPONSE2 gene is required for auxin synthesis and diverse aspects of plant development. *Plant Physiol* 151, 168-179

He, W., Brumos, J., Li, H., Ji, Y., Ke, M., Gong, X., Zeng, Q., Li, W., Zhang, X., An, F., Wen, X., Li, P., Chu, J., Sun, X., Yan, C., Yan, N., Xie, D. Y., Raikhel, N., Yang, Z., Stepanova, A. N., Alonso, J. M., and Guo, H. (2011) A small-molecule screen identifies L-kynurenine as a competitive inhibitor of TAA1/TAR activity in ethylene-directed auxin biosynthesis and root growth in Arabidopsis. *Plant Cell* 23, 3944-3960

Nishimura, T., Hayashi, K., Suzuki, H., Gyohda, A., Takaoka, C., Sakaguchi, Y., Matsumoto, S., Kasahara, H., Sakai, T., Kato, J., Kamiya, Y., and Koshiba, T. (2014) Yucasin is a potent inhibitor of YUCCA, a key enzyme in auxin biosynthesis. *Plant J* 77, 352-366

Korasick, D. A., Enders, T. A., and Strader, L. C. (2013) Auxin biosynthesis and storage forms. *J Exp Bot*, 64(69):2541-2555

Staswick, P. E., Serban, B., Rowe, M., Tiryaki, I., Maldonado, M. T., Maldonado, M. C., and Suza, W. (2005) Characterization of an Arabidopsis enzyme family that conjugates amino acids to indole-3-acetic acid. *Plant Cell* 17, 616-627

Qin, G., Gu, H., Zhao, Y., Ma, Z., Shi, G., Yang, Y., Pichersky, E., Chen, H., Liu, M., Chen, Z., and Qu, L. J. (2005) An indole-3-acetic acid carboxyl methyltransferase regulates *Arabidopsis* leaf development. *Plant Cell* 17, 2693-2704

Jackson, R. G., Lim, E. K., Li, Y., Kowalczyk, M., Sandberg, G., Hoggett, J., Ashford, D. A., and Bowles, D. J. (2001) Identification and biochemical characterization of an Arabidopsis indole-3-acetic acid glucosyltransferase. *J Biol Chem* 276, 4350-4356

Tognetti, V. B., Van Aken, O., Morreel, K., Vandenbroucke, K., van de Cotte, B., De Clercq, I., Chiwocha, S., Fenske, R., Prinsen, E., Boerjan, W., Genty, B., Stubbs, K. A., Inze, D., and Van Breusegem, F. (2010) Perturbation of indole-3-butyric acid homeostasis by the UDP-glucosyltransferase UGT74E2 modulates Arabidopsis architecture and water stress tolerance. *Plant Cell* 22, 2660-2679

Zhao, Z., Zhang, Y., Liu, X., Zhang, X., Liu, S., Yu, X., Ren, Y., Zheng, X., Zhou, K., Jiang, L., Guo, X., Gai, Y., Wu, C., Zhai, H., Wang, H., and Wan, J. (2013) A role for a dioxygenase in auxin metabolism and reproductive development in rice. *Dev Cell* 27, 113-122

Pěnčík, A., Simonovik, B., Petersson, S. V., Henyková, E., Simon, S., Greenham, K., Zhang, Y., Kowalczyk, M., Estelle, M., Zažímalová, E., Novák, O., Sandberg, G., and Ljung, K. (2013) Regulation of auxin homeostasis and gradients in *Arabidopsis* roots through the formation of the indole-3-acetic acid catabolite 2-oxindole-3-acetic acid. *Plant Cell* 25, 3858-3870

Tanaka, K., Hayashi, K., Natsume, M., Kamiya, Y., Sakakibara, H., Kawaide, H., and Kasahara, H. (2014) UGT74D1 catalyzes the glucosylation of 2-oxindole-3-acetic acid in the auxin metabolic pathway in *Arabidopsis*. *Plant Cell Physiol* 55, 218-228

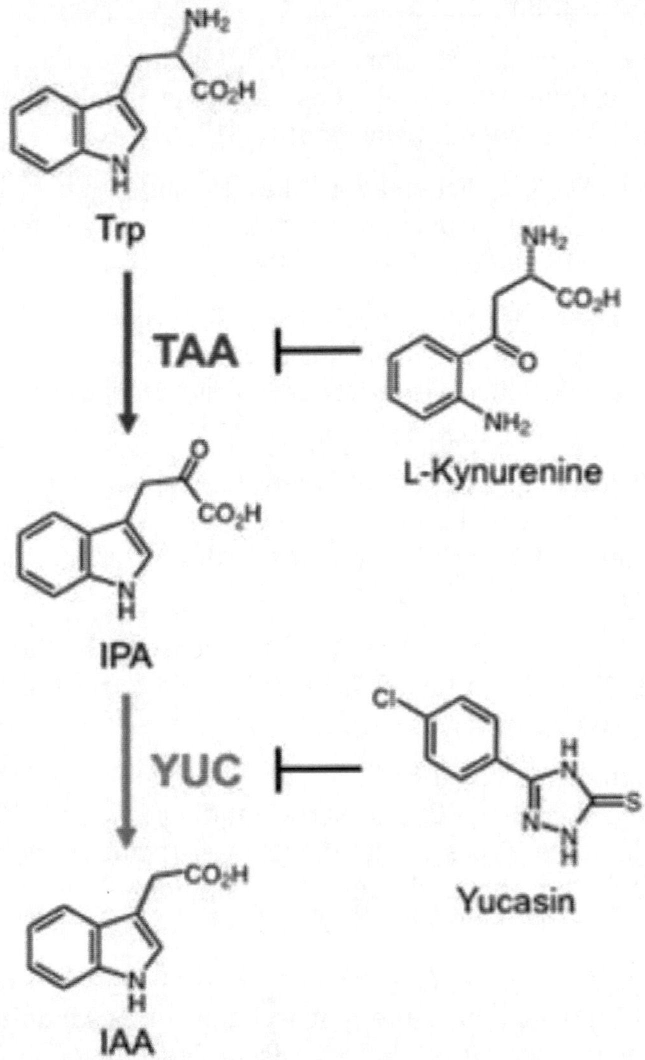

Figure 1. Main IAA biosynthesis pathway in plants and the structures of TAA and YUC inhibitors.

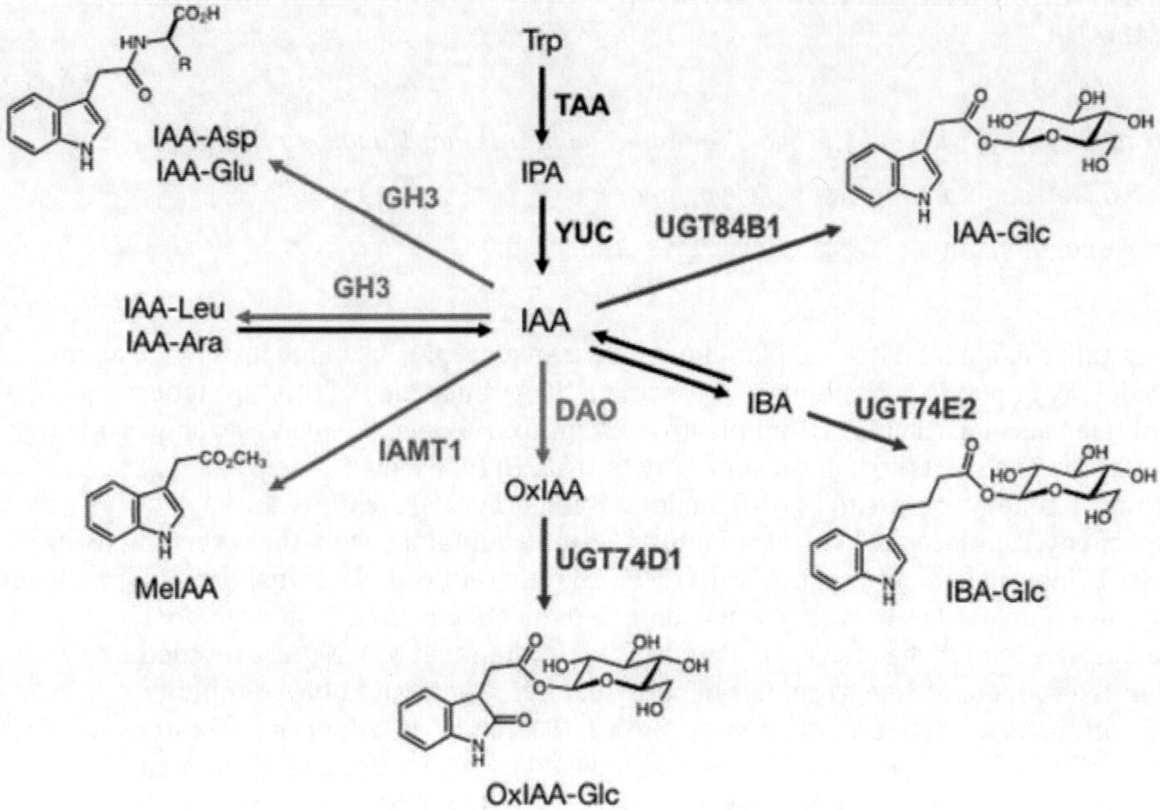

Figure 2. Proposed IAA inactivation pathways in plants.

A 2,4-D ANALOG EXHIBITS AN INHIBITION ON AUXINS INFLUX IN *ARABIDOPSIS THALIANA*

Hiromi Suzuki[1,2], Naoyuki Matano[1], Takeshi Nishimura[1] and Tomokazu Koshiba[1]

[1]*Tokyo Metropolitan University, Tokyo, Japan*

[2]*Corresponding author: suzuki-hiromi1@ed.tmu.ac.jp*

The studies using inhibitors of plant hormone transport give us many insights of auxin physiology in plants. 1-Naphtoxyacetic acid (1-NOA), an analog of the synthetic auxin 1-N-naphtalene acetic acid (NAA), inhibits the IAA influx. However, 1-NOA also shows auxin activity because of its structural similarity to NAA. Here, we identified a 2,4-dichlorophenoxyacetic acid (2,4-D) analog, "7-B3; ethyl 2-[(2-chloro-4-nitrophenyl)thio]acetate", can also inhibit IAA influx obtained from the screening using maize coleoptile. At more than 300 µM, 7-B3 slightly reduced IAA transport and gravitropic response of maize coleoptiles. We also detected the effects of 7-B3 on *Arabidopsis* seedlings. Although the effect of 7-B3 was weaker than that of 1-NOA, it rescued 2,4-D-inhibited root elongation as similar to non-treatment root, but not NAA-inhibited elongation. As for *DR5::GUS* expression, both 1-NOA and 7-B3 reduced its expression induced by IAA and 2,4-D, but did not that induced by NAA. Interestingly at high concentration, 1-NOA exhibited auxin activity, but 7-B3 did not. Furthermore, 7-B3 inhibited apical hook formation in etiolated seedlings more effectively than 1-NOA. Therefore, we concluded that 7-B3 could be an inhibitor of IAA influx almost without effect on IAA efflux or auxin activity.

P450S IN TRITERPENOID BIOSYNTHESIS: DIVERSITY AND THEIR APPLICATION TO COMBINATORIAL BIOSYNTHESIS

Toshiya Muranaka[1,2]

[1]Osaka University, Japan

[2]Corresponding author: muranaka@bio.eng.osaka-u.ac.jp

Plants produce a wide variety of specialized (secondary) metabolites, with which they interact in various environmental conditions for survival. Cytochrome P450s have a central function to enhance the diversity of the chemicals. The study focused on the diversity of P450s in (tri) terpenoid biosynthesis and their application to combinatorial biosynthesis has been performed in my research group. A strategy combining a homology-based approach, gene coexpression analysis, and combinatorial biosynthesis with heterologous expression in yeast was successful in identifying enzymes involved in triterpenoid biosynthesis and also in generating natural and rare triterpenoids that do not accumulate in planta. Using this strategy it is possible to construct a natural/unnatural triterpenoid library for further application to discover new type of plant growth regulators. The next steps are then to increase product yields as well as to diversify triterpenoids into novel synthetic entities with improved biological activities by combining enzymes from different sources.

OVEREXPRESSION OF THE bZIP TRANSCRIPTION FACTOR OSBZIP79 SUPPRESSES THE PRODUCTION OF DITERPENOID PHYTOALEXIN IN RICE CELLS

Koji Miyamoto[1,4], Yoko Nishizawa[2], Eiichi Minami[2], Hideaki Nojiri[3], Hisakazu Yamane[1], Kazunori Okada[3]

[1]Department of Biosciences, Teikyo University, 1-1 Toyosatodai, Utsunomiya, Tochigi 320-8551, Japan

[2]Disease Resistant Crops Research Unit, GMO Research Center, National Institute of Agrobiological Sciences, 2-1-2 Kannondai, Tsukuba, Ibaraki 305-8602, Japan

[3]Biotechnology Research Center, The University of Tokyo, 1-1-1 Yayoi, Bunkyo-ku, Tokyo 113-8657, Japan

[4]Corresponding author: miyamoto@nasu.bio.teikyo-u.ac.jp

When plants are attacked by pathogenic microorganisms, they respond with a variety of defensive reactions, such as the production of antimicrobial secondary metabolites phytoalexins. Momilactones and phytocassanes, major diterpenoid phytoalexins in rice, are inductively synthesized from geranylgeranyl diphosphate derived from methylerythritol phosphate (MEP) pathway. We previously reported that OsTGAP1, a chitin oligosaccharide elicitor-inducible bZIP transcription factor, is involved in the regulation of the inductive expression of biosynthetic genes, including the MEP pathway genes, responsible for diterpenoid phytoalexins production. To obtain a clue to elucidate regulatory mechanisms of OsTGAP1-mediated production of diterpenoid phytoalexins in rice cells, we performed yeast two-hybrid screening to isolate OsTGAP1–interacting proteins. Among the candidates of OsTGAP1-interacting proteins, we focused a TGA factor, OsbZIP79, and examined its physical interaction with OsTGAP1 and contribution to the phytoalexins productions. *In vitro* pull-down assay demonstrated that OsTGAP1 and OsbZIP79 exhibited heterodimeric as well as homodimeric interaction. Intriguingly, despite evident transactivation activity of the OsbZIP79 in a transient reporter assay, overexpression of *OsbZIP79* resulted in suppression of the inductive expression of diterpenoid phytoalexin biosynthetic genes, and thus caused decrease in the accumulation of phytoalexin in rice cells. These results suggest that OsbZIP79 functions as a negative regulator of the phytoalexin production triggered by the chitin oligosaccharide elicitor in rice cells.

BLUE-LIGHT IRRADIATION UP-REGULATES THE *ENT*-KAURENE SYNTHASE GENE AND AFFECTS THE AVOIDANCE RESPONSE OF PROTONEMAL GROWTH IN *PHYSCOMITRELLA PATENS*

Sho Miyazaki[1,4], Masatoshi Nakajima[2], Masahiro Natsume[3], Hiroshi Kawaide[3] and Hikaru Toyoshima[3]

[1]The University of Tokyo, Bunkyo-ku, Tokyo, Japan; Tokyo University of Agriculture and Technology, Fuchu, Tokyo, Japan

[2]The University of Tokyo, Bunkyo-ku, Tokyo, Japan

[3]Tokyo University of Agriculture and Technology, Fuchu, Tokyo, Japan

[4]Corresponding author: miyazaki@pgr1.ch.a.u-tokyo.ac.jp

Gibberellins (GAs) are a group of diterpene-type plant hormones biosynthesized from *ent*-kaurenoic acid *via ent*-kaurene. While the moss *Physcomitrella patens* has part of the GA biosynthetic pathway, from geranylgeranyl diphosphate to *ent*-kaurenoic acid, no GA is found in this species. Caulonemal differentiation in a *P. patens* mutant with a disrupted bifunctional *ent*-copalyl diphosphate synthase/*ent*-kaurene synthase (PpCPS/KS) gene is suppressed under red light, and is recovered by application of *ent*-kaurene and *ent*-kaurenoic acid. This indicates that derivatives of *ent*-kaurenoic acid, not GAs, might act as endogenous developmental regulators. In this research, we show unique protonemal growth responses of *P. patens* under unilateral blue light, and these regulators were involved in the responses. When protonemata of the wild type were incubated under blue light, the chloronemal filaments grew in the opposite direction to the light source. Although this avoidance response was not observed in the *ent*-kaurene deficient mutant, chloronemal growth toward a blue-light source in the mutant was suppressed by application of *ent*-kaurenoic acid, and the growth was rescued to that in the wild type. Expression analysis of the *PpCPS/KS* gene showed that the mRNA level under blue light was rapidly increased and was five times higher than under red light. These results suggest that hormonal diterpene compounds derived from *ent*-kaurenoic acid *via ent*-kaurene are responsible for regulation of blue-light avoidance in *P. patens*.

DOES THE BRASSINOSTEROID SIGNAL PATHWAY IN PHOTOMORPHOGENESIS OVERLAP WITH THE GRAVITROPIC RESPONSE CAUSED BY AUXIN?

Masatoshi Nakajima[1,4], Tadao Asami[1], Naiyanate Jaroensanti[2], Jung-Min Yoon[2], Yuji Nakai[2], Ikuya Shirai[2], Masato Otani[2], Seung-Hyun Park[2], and Ken-ichiro Hayashi[3]

[1]The University of Tokyo, Tokyo, Japan, CREST, Japan Science and Technology Agency, Japan

[2]The University of Tokyo, Tokyo, Japan

[3]Okayama Univ. of Science, Okayama, Japan

[4]Corresponding author: anakajm@mail.ecc.u-tokyo.ac.jp

Brassinosteroid (BR) and auxin co-regulate plant growth in a process termed cross-talking. Based on the assumption that their signal transductions are partially shared, inhibitory chemicals for both signal transductions were screened from a commercially-available library. A chemical designated as NJ15 diminished the growth promotion of both adzuki bean epicotyls and Arabidopsis seedlings, by either the application of BR or auxin. To understand its target site(s), bioassays with a high dependence on either the signal transduction of BR (BR-signaling) or of auxin (AX-signaling), were performed. NJ15 inhibited photomorphogenesis of Arabidopsis seedlings grown in the dark, which mainly depends on BR-signaling, while NJ15 also inhibited their gravitropic responses mainly depending on AX-signaling. On the study for the structure-activity relationships of NJ15 analogues, they showed strong correlations on the inhibitory profiles between BR- and AX-signalings. These correlations imply that NJ15 targets the downstream pathway after the integration of BR- and AX-signals.

BIOLOGICAL ROLES OF SAKURANETIN, A FLAVONOID SPECIALIZED METABOLITE INDUCTIVELY PRODUCED IN RICE

Satoshi Ogawa[1,7], Hideaki Nojiri[1], Kazunori Okada[1], Chie Yoshiga[1,2], Gen-ichiro Arimura[2], Takafumi Shimizu[3], Koji Miyamoto[4], Koichi Hamada[5], Yoko Nishizawa[6] and Eiichi Minami[6],

[1]The University of Tokyo, Tokyo, Japan

[2]Tokyo University of Science, Tokyo, Japan

[3]RIKEN Center for Sustainable Resource Science, Kanagawa, Japan

[4]Teikyo University, Tochigi, Japan

[5]Showa Pharmaceutical University, Kanagawa, Japan

[6]Disease Resistant Crops Research Unit, GMO Research Center, National Institute of Agrobiological Sciences, Ibaraki, Japan

[7]Corresponding author: 7624881517@mail.ecc.u-tokyo.ac.jp

Sakuranetin, the major flavonoid phytoalexin in rice, can be induced by ultraviolet (UV) irradiation, treatment with $CuCl_2$ or jasmonic acid (JA), or phytopathogenic infection. In addition to the biological role of sakuranetin on disease resistance in rice, its broad bioactivities have recently been described. Results from these studies have shown that sakuranetin is a useful compound as a plant antibiotic and a potential pharmaceutical agent. Sakuranetin is biosynthesized from naringenin, a precursor of sakuranetin, by naringenin 7-*O*-methyltransferase (NOMT), but the relevant gene had not been identified in rice. Recently, we identified the *OsNOMT* gene, which is involved in the final step of sakuranetin biosynthesis in rice[1)]. Gene expression was induced by treatment with JA and $CuCl_2$ in rice leaves prior to sakuranetin accumulation. Furthermore, an *osjar1* mutant, which is defective in synthesis of an active jasmonate, JA-isoleucine, was shown to be impaired in both JA- and $CuCl_2$-induced expression of the *OsNOMT* genes and sakuranetin production, suggesting JA-mediated control of sakuranetin synthesis[2)]. Identification of the *OsNOMT* gene enables potential production of large amounts of sakuranetin through transgenic rice. Therefore, we generated transgenic rice plants overexpressing the *OsNOMT* gene and productivity of sakuranetin was examined. Besides, *OsNOMT*-knockdown plants generated by RNAi approach were successfully obtained as a sakuranetin-deficient rice. By using these *OsNOMT*-modified plants, we have been examining the contribution of sakuranetin on resistance to biotic and abiotic stresses such as blast fungus infection, UV irradiation, and others. We discuss the biological significance of sakuranetin in rice and potential utilization of this specialized metabolite as a pharmacological agent as well.

LITERATURE CITED

Shimizu *et al.* (2012) 287:19315-19325, *J. Biol. Chem.*

Shimizu *et al.* (2013) 77:1556-1564, *Biosci. Biotechnol. Biochem.*

GROWTH CONTROL OF LEAFY VEGETABLES WITH S-ABSCISIC ACID (S-ABA) FOR IMPROVED QUALITY AND HARVEST MANAGEMENT

Jozsef Racsko[1,2], F. Marmor[1], Peter D. Petracek[1] and J. Pienaar[1]

[1]Valent BioSciences Corporation, Libertyville, IL, USA

[2]Corresponding author: jracs@valent.com

Consumption and therefore production of leafy green vegetables and salad mixes have been increasingly popular. Grower price of leafy vegetables, e.g., spinach, for fresh consumption is primarily determined by leaf size. Higher commercial value is associated with smaller leaf size (<3 inches), known as baby leaf. Larger leaf size is of reduced commercial value and prone to mechanical injury. Optimum harvest time is very narrow in leafy vegetables due to their fast growth (23-38 day production cycle). Warm/hot growing temperatures that is characteristic of the major growing areas of the US (i.e., CA, AZ) often speeds up maturity to faster than planned. There is a strong need to hold leaf size for several days to keep value (i.e., in baby leaf stage) and time harvest. Recently, S-ABA has been proven to effectively control leaf growth and keep leaf size of spinach at high commercial value level for 3-5 days, without side effects. There is an excellent crop safety with other leafy green vegetables with spray applications of S-ABA up to 2,000 ppm concentration (e.g., red leaf lettuce, baby green Romaine, Lolla rosa, mizuna, tango, beet tops, Swiss chard, parsley). This study gives a report on the potentials of S-ABA use in leafy green vegetables from a series of field trials conducted under commercial production conditions. Under the trade name ConTego™ SL, S-ABA has recently received federal registration, and may become an important tool for leafy green vegetable producers to control leaf size, time harvest and ultimately improve grower profitability.

EFFECT OF THE PREPARATION FROM *AILANTHUS ALTISSIMA,* CE AT VARIOUS STAGES OF THE SPIDER MITE ONTOGENESIS

Dzhalil Chakaeva[1], T. Chermenskaya[1] and Ulanbek Salaidinov[1,2]

[1]Bio-Kg, Dushanbinskaya 10-38, Bishkek, Kyrgyzstan 720000

[2]Corresponding author: ulan.salaidinov@gmail.com

Pest processing at different stages of ontogenesis was made for determining the period of peak sensitivity of the spider mite to the most promising preparation from *A. altissima* and for studying the mechanism of its action. Experimental results clearly demonstrate the ability of the preparation to produce inhibitory action on the reproductive system of female mites, which is shown in 3 times lowering of the fertility of processed females and in veracious lowering of larvae birth. Only insignificant changes in mite population were noticed under the influence of the extract on the specimens of embryonal stage (diurnal eggs). Processing foliage with young larvae mite induces larvae death mainly in a day, i.e. the specimens which were under direct treatment. Later on, the development goes like in the control variant. The data on biological effectiveness demonstrate clear dependence of this characteristic on the stage of mite ontogenesis being processed.

DETERMINATION OF WATER-USE EFFICIENCY IN PGR-TREATED WINTER WHEAT AT DIFFERENT LEVELS OF WATER SUPPLY

Wilhelm Rademacher[1,2,3]

[1]*BASF Global Research Crop Protection, 67117 Limburgerhof, Germany*

[2]*Corresponding author: radifax@t-online.de*

[3]*Current address: Austrasse 1, 67117 Limburgerhof, Germany*

INTRODUCTION

The plant growth regulator Medax ® Top (trade name in the UK and in Ireland: Canopy®) contains mepiquat chloride and prohexadione-Ca (Fig. 1). Both active ingredients inhibit gibberellin biosynthesis, albeit at different sites of the pathway (Rademacher, 2000). Medax ® Top reduces shoot elongation and is used to lower the risk of lodging in intense cereal production. It has often been claimed that anti-lodging agents lead to less water consumption and increase water-use efficiency (WUE) as a result of reduced leaf surface and a more compact shoot growth. In order to help answer this question, a trial with wheat was conducted in 2007 in BASF's vegetation hall in Limburgerhof. Here, plants can be grown to full maturity under field-like conditions on a rotating belt with protection against rain and wind given by a retractable transparent roof.

MATERIALS AND METHODS

Spring wheat (cv. 'Triso') was grown in 5 l Mitscherlich pots (12 plants per pot) in a light soil (loamy sand) containing adequate amounts of nutrients. Automatic irrigation (up to three times a day) was adjusted to 60% (regular watering) or 30% (moderate water shortage) of the soil's field capacity (FC). Plants were treated two times with, each, 750 ml/ha of Medax ® Top at growth stages 31 (first node at least 1 cm above tillering node) and 39 BBCH (flag leaf unrolled), which was on May 2 and May 25, respectively. 750 g/ha of ammonium sulfate was added at each application as a water conditioner. Each control and treatment at the different levels of water supply consisted of four parallels. Water consumption was continuously monitored from April 26 until July 18, when the plants were ready for harvest. Plant surface temperature was recorded on June 6 with a FLIR ThermaCAM SC3000 infrared camera.

RESULTS AND DISCUSSION

The obtained results are shown in Table 1: Treatments with Medax ® Top led to significant reductions of shoot length. At regular water supply, this was correlated with an increase of seed yield of almost 11%. Most likely, assimilates no longer needed for shoot growth were re-directed into the developing seeds. However, Medax ® Top applied to drought-stressed plants had a significant negative impact on seed yield. Obviously, the reduction of shoot growth resulting from, both, PGR treatment and water stress had become too intense. The only scenario with an improved WUE consisted of keeping plants constantly under a modest drought stress. Although there was no reduction of seed yield in comparison to regularly watered plants, only 587 ml of water were required to produce 1 g of seeds. In contrast, regularly watered plants consumed 718 ml of water per 1 g of seeds.

The results of this investigation do not support the view that reducing shoot growth with anti- lodging agents leads to lowered water consumption and improved WUE. Infrared pictures (data not shown) indicate that the compacted parts of PGR-treated and regularly watered plant are lower in temperature than equivalent parts of untreated plants. This indicates that they are more actively transpiring water (and assimilating CO2) than untreated plants in this area. In contrast, plants cultivated with reduced supply of water do not display such type of a difference. Improving WUE without having a decrease in seed yield just by lowering water supply also demonstrates that plants are "wasting" water, when there is ample supply. It also indicates that a more successful approach for crop production at limited water availability should rather be via regulation of stomatal aperture. Positive results on WUE in seed production in barley by using abscisic acid and an analog of this plant hormone are available (Rademacher et al., 1989).

Literature Cited

Rademacher, W. (2000). Growth retardants: Effects on gibberellin biosynthesis and other metabolic pathways. Annual Review of Plant Physiology and Plant Molecular Biology 51, 501- 531.

Rademacher, W., Maisch, R., Liessegang, J. and Jung, J. (1989). New synthetic analogues of abscisic acid: Their influence on water consumption and yield formation in crop plants. In: Structural and Functional Responses to Environmental Stresses, Kreeb, K.H., Richter, H., Hinckley, M. (eds.), SPB Academic Publishing bv, The Hague, pp. 147-154.

Acknowledgement

Technical support given by Wolfgang Weigelt, Pilar Puente and Ulrich Seitz (all BASF Limburgerhof) is gratefully acknowledged.

Table 1. Effects of Medax® Top on shoot growth, water consumption, seed yield, and water-use efficiency in spring wheat cv. Trisò.

Treatment/Water Supply	Final shoot length (cm)	Water consumption (mL/pot)	Seed yield (g/pot)	Water-use efficiency (mL/g of seeds)
Control/60% FC	81.8	28,500	39.7	718
Medax® Top/60% FC	60.56	29,925	42.9	698
Control/30% FC	70.3	23,085	39.3	587
Medax® Top/30% FC	54.0	21,945	30.6	717

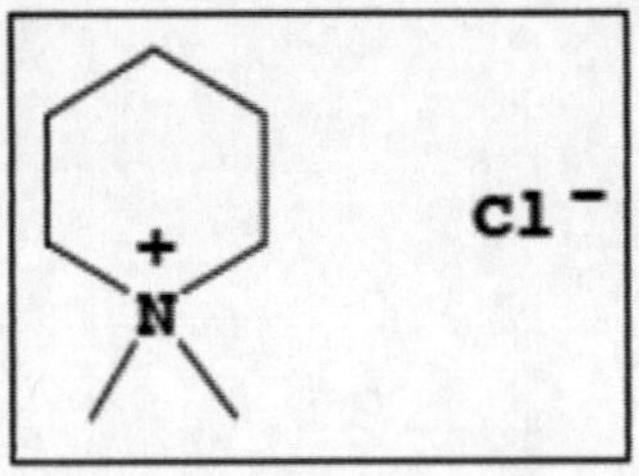

Mepiquat chloride

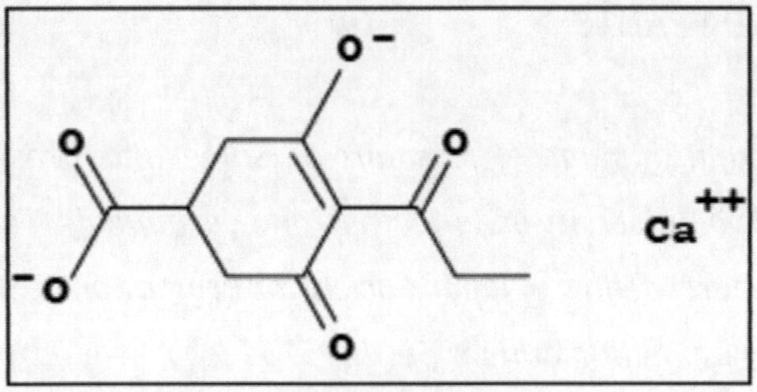

Prohexadione-Ca

Figure 1. Medax® Top (Canopy®) is an SC formulation with 300 g/L of mepiquat chloride and 50 g/L of prohexadione-Ca.

SUBSTRATE RECOGNITION-PRODUCT DIVERSITY OF *ENT*-KAURENE SYNTHASES IN LAND PLANTS

Manami Shimane[1,2,3], Masahiro Natsume[1] and Hiroshi Kawaide[1]

[1]Tokyo University of Agriculture and Technology, Tokyo, Japan

[2]Research fellow of Japan Society for Promotion of Science (JSPS)

[3]Corresponding author: 50012951012@st.tuat.ac.jp

ent-Kaurene, a diterpene precursor of gibberellins, is biosynthesized from geranylgeranyl diphosphate (GGDP) via *ent*-copalyl diphosphate (*ent*-CDP). In flowering plants, monofunctional *ent*-CDP synthase (*ent*-CPS) and *ent*-kaurene synthase (KS) catalyze each cyclization reactions from GGDP to *ent*-CDP, from *ent*-CDP to *ent*-kaurene, respectively. Non-vascular plants have bifunctional CPS/KS which catalyzes both reactions of *ent*-CPS and KS in single polypeptide. To get more insight into the evolution of land plant gibberellin biosynthesis, we focused on diterpene cyclases of lycophyte *Selaginella moellendorffii*.

We characterized two diterpene cyclases of *S. moellendorffii*, SmKS and SmDTC3, *in vitro*. SmKS is phylogenetically close to known KSs of flowering plants and identified as monofunctional KS as expected. SmDTC3 also showed only KS-like activity although it shares 76% amino acid sequence identity with bifunctional miltiradiene synthase of *S. moellendorffii* (Sugai *et al., J. Biol. Chem.,* 2011). SmDTC3 converted *ent*-CDP, normal CDP (also known as (+)-CDP) and *syn*-CDP to *ent*- 16a-hydroxykaurane, sandaracopimaradiene and multiple diterpene products, respectively.

Interestingly, SmKS also catalyzed similar reactions when normal CDP or *syn*-CDP was subjected instead of *ent*-CDP substrate: SmKS synthesized sandaracopimaradiene from normal CDP and multiple hydrocarbons from *syn*-CDP. Therefore, substrate recognition of KSs from other plants was investigated. The moss *Physcomitrella patens* CPS/KS (PpCPS/KS) and lettuce KS (LsKS) can utilize all three CDP stereoisomers as substrate like SmDTC3 and SmKS, but rice KS has strict substrate specificity to *ent*-CDP. These results are summarized in Table. Semi-quantitative analysis indicated that SmDTC3, SmKS and PpCPS/KS efficiently convert normal CDP and *syn*-CDP whereas LsKS is highly selective to *ent*-CDP rather than normal and *syn*-CDP. Our studies imply that ancient KS having low substrate specificity has evolved to be specific for *ent*-CDP to the biosynthesis of gibberellin.

Table 1. Cyclization products from three CDP stereoisomers by diterpene cyclases in the present study.

Enzyme	Substrate and cyclization products		
	ent-CDP	normal CDP	*syn*-CDP
SmKS	*ent*-Kaurene	Sandaracopimaradiene	Multiple hydrocarbons
SmDTC3	*ent*-16α-Hydroxykaurene + hydrocarbons (minor)	Sandaracopimaradiene	Two diterpene alcohols + hydrocarbons (minor)
PpΔcps/KS[z]	*ent*-16α-Hydroxykaurene + *ent*-Kaurene (minor)	Multiple hydrocarbons	Two hydrocarbons
LsKS	*ent*-Kaurene	Sandaracopimaradiene + one hydrocarbon (minor)	Multiple hydrocarbons
OsKS1	*ent*-Kaurene	Note detected	Not detected

[z] PpΔcps/KS is the mutant enzyme of PpCPS/KS which shows only KS activity and loss of *ent*-CPS activity because of point mutations in CPS active site.

EFFECT OF STRIGOLACTONE ON TAPROOT GROWTH IN RADISH (*RAPHANUS SATIVUS* L.)

Ikuo Takahashi[1,2], Kosuke Fukui[1], Hidemitsu Nakamura[1] and Tadao Asami[1]

[1]The University of Tokyo, Japan

[2]Corresponding author: itakahashi@pgr1.ch.a.u-tokyo.ac.jp

Radish is a Brassicaceae root vegetable produced throughout the world. The Far Eastern Asian countries including Japan are the major consumption area. Radish taproot growth is dependent on exo- and endogenous signals such as light, nutrition and plant hormone. Exogenously applied cytokinin and auxin promote taproot growth, however, contribution of other plant hormones to the growth of taproot has not been well investigated. Strigolactones (SLs) are a group of terpenoid lactone-type plant hormones derived from carotenoid, which was identified as germination inducers of parasitic weeds and as a trigger for interactions of plant roots with mycorrhizal fungi. In addition, SLs have been shown to play roles in many different aspects in plant architecture, such as shoot branching, primary root growth, and photo-morphogenesis. Here we demonstrate the effect of SLs on radish growth. We found that application of a typical synthetic SL mimic, GR24, to soil-grown radish could increase fresh weight and diameter in taproot. In this study, *in vitro* culture system has been used to investigate the growth-promoting effect of SLs in detail. Application of GR24 (1-20 μM) to an *in vitro* cultured radish also resulted in an increase in radial growth in taproot. Application of plant-specific SL mimic, 4-Br debranone (4-BD), was more effective in this promoting effect at the concentrations of 1-20 μM. Anatomical study showed that radial growth resulted from expansion of central cylinder. Taken together, it is suggested that SLs contribute to cell division and expansion during the growth of radish taproot. In addition, result of screening to find more effective 4-BD derivatives is also reported.

BROWNING MECHANISM ON THE ROOTS OF RICE PLANT INFESTED BY RICE ROOT APHID, *RHOPALOSIPHUM RUFIABDOMINALIS*, AND EFFECTS OF SALICYLIC ACID

Shinichi Tebayashi[1,7], Chisato Sano[1], Shinji Ueda[1], Akira, Oikawa[2,3], Ryosuke Sasaki[3], Hideaki Maseda[4], Kazuki Saito[3,5], and Atsushi Ishihara[6]

[1]Kochi University, Japan

[2]Yamagata University

[3]RIKEN PSC, Japan

[4]University Tokushima, Japan

[5]Chiba University, Japan

Tottori University, Japan

[7]Corresponding author: tebayasi@kochi-u.ac.jp

A browning on rice plant foliages caused by attacking pathogens is one of important defensive systems. Ishihara *et al*. (2004) had demonstrated that the browning on rice plant leaf acted as a defense against pathogens and the browning material would be accumulated by polymerization of serotonin. However the role and the mechanism of a browning on a rice plant root attacked by insects is unclear. Material and Method. Insects and plants: Rice sterilized seeds were sown on a paper towel, and seedling were cultivated on a paper towel with distillate water at 25 ± 3 °C(16L: 8D). A root of rice seeding at 4 days old was infested by a winged aphid and they were kept under the same condition. A length of the root, a color of the root, and number of the aphid was measured.

Chemical analysis: Rice roots (ca. 20 mg) were immersed in 200 μL MeOH, and homogenized. The extract was analyzed by CE-TOFMS, which was performed using an Agilent CE capillary electrophoresis system (Agilent Technologies, Waldbronn, Germany)

Quantitative RT-PCR: Total RNA was extracted from the rice root (200 mg) using an RNeasy Plant Mini Kit (Qiagen). Single-strand cDNA was synthesized from the total RNA by reverse transcription using a ReverTra Ace Kit (TOYOBO). Quantitative real-time PCR was performed using a Step One Plus (ABI) with specific primer sets and a real-time PCR master mix (Thunderbird, TOYOBO).

Peroxidase activities: Peroxidase (POX) activity was measured by using a crude enzyme solution and serotonin as substrate. The crude enzyme solution was prepared from roots (10 mg) and 1.5 mL of 50 mM K-Pi buffer (pH 6.9). The reaction mixture, which consist of 100 mM Mcllvain buffer (pH 6.0), 7 mM H_2O_2, and 4.8 mM serotonin, was incubated at 30 °C for 5 min. The reaction was stopped by 25 μL of 6 M HCl and 225μL of MeOH, and the reaction mixture was analyzed by HPLC.

Result and discussion. The browning was induced by an infestation of aphids, and the aphids avoided the deep browning area on the roots. Thus it was thought that rice roots could induce any defensive system with a browning, when they were attacked by pests. The metabolome analyses of roots extract revealed that an infestation of an aphid on rice roots induced an accumulation of serotonin and tryptamine. Furthermore, a high concentration of serotonin inhibited the survival of nymph aphids on a rearing test using an artificial diet. Therefore, it was concluded that rice plants could induce a chemical defense system against insect pests. Next, an accumulation mechanism of serotonin on a root was also investigated. When a root was attacked by aphids, the transcript of serotonin biosynthesis-related genes (*Anthranil synthase* (*AS*), *Tryptophan synthase* (*TS*), *Tryptophan decarboxylase* (*TDC*), and *Tryptamine-5-hydroxylase* (*T5H*)) was induced. On the other hand, an activation of POX activity, which metabolizes serotonin, is delayed than the expression of T5H, which biosynthesis of serotonin. Thus, this deviation of the expression timings would induce the accumulation of serotonin. Finally, an effect of salicylate of browning on a root was examined. The expression of salicylate biosynthesis-related gene, *Isochorismate mutase* (*ICM*), was increased by an infestation of aphids. Furthermore, when the roots were infected by aphids, the root treated with salicylate was browning deeper than the control root early on. Then it was assumed that the browning on rice roots was regulated by salicylate.

AUXIN AND EXPRESSION OF *TAC1* AND *MAX1-4* IN DIFFERENT GROWTH HABITS OF PEACH

Thomas Tworkoski[1,2], Kevin Webb[1] and Ann Callahan[1]

[1]USDA, ARS, Appalachian Fruit Research Station, 2217 Wiltshire Rd., Kearneysville, WV 25430 USA

[2]Corresponding author: tom.tworkoski@ars.usda.gov

Branch orientation and distribution are fundamental aspects of tree architecture that influence orchard design and management. The goal of the study was to identify genetically modifiable regulatory processes or those that can be managed culturally to customize tree architecture. Peach (*Prunus persica* L. (Batch)) trees with three different branching genotypes were evaluated: nearly vertical branches (pillar), less vertical and more spreading branches (upright), and least vertical branches (standard). Auxin concentrations and gene expression of key branching enzymes in herbaceous species, *MAX1, 2, 3,* and *4*, were determined. Also, expression of the gene, *TAC1*, that is associated with branch angle and serves as a marker for pillar peach trees was measured. Shoots and roots of peach trees in the field and greenhouse were studied during periods of growth when bud break and branch spatial orientation develop. Endogenous auxin concentrations were determined by mass spectrometry and gene expression was relatively quantified with real-time PCR. Expression of TAC1 in shoots was greatest in standard and least in pillar trees. This supports previous work where unexpressed TAC1 was associated with narrow branch angles as found in pillar trees. Auxin and the branch-related genes, *MAX1-4,* were dissimilar between standard and pillar or upright growth habits and expression changed during the growing season. Gene expression of *MAX1-4* was higher in roots than stems but it did not differ among the roots or rootstocks of the different growth habits. The current work indicates that in stems, auxin and *MAX1-4* genes may influence regulatory processes that affect growth and development of peach trees with different growth habits. In addition to breeding, new plant growth regulators that affect the modes of action of root- originating signals, possibly strigolactone, may provide new cultural tools for managing tree growth and development.

AUXIN POLAR TRANSPORT IS ESSENTIAL FOR GRAVIRESPONSE OF ETIOLATED PEA SEEDLINGS

Junichi Ueda[1], Motoshi Kamada[2], Mariko Oka[3], Kensuke Miyamoto[4], Eiji Uheda[1], Akira Higashibata[5]

[1]Graduate School of Science, Osaka Prefecture University

[2]Future Development Division, Advanced Engineering Service Corporation

[3]Faculty of Agriculture, Tottori University

[4]Faculty of Liberal Arts and Sciences, Osaka Prefecture University

[5]Tsukuba Space Center, Japan Aerospace Exploration Agency

[6]Corresponding author: ueda@b.s.osakafu-u.ac.jp

The purpose of this study is to verify the hypothesis that auxin polar transport is essential for graviresponse in plants, focused on the mode of action of auxin polar transport on graviresponse of etiolated Alaska pea seedlings. Automorphogenesis of etiolated Alaska pea (*Pisum sativum* L.) seedlings together with reduced auxin polar transport in epicotyls was observed in true microgravity conditions in space (BRIC-AUX on STS-95 in 1998). On Earth, a potent inhibitor of auxin polar transport, 2,3,5-triiodobenzoic acid (TIBA), substantially induced automorphogenesis-like growth and development in etiolated Alaska pea seedlings. Similar powerful inhibitors of auxin polar transport, *N*-(1-naphthyl)phthalamic acid (NPA) and 9-hydroxyfluorene-9-carboxylic acid (HFCA), were found to phenocopy automorphogenesis-like epicotyl bending in etiolated Alaska pea seedlings. However, an inhibitor of auxin action, *p*-chlorophenoxyisobutylic acid (PCIB), had little effect. Auxin polar transport in the proximal side of the 1st internodes to the cotyledons, but not in the 2nd internodes, in etiolated Alaska pea seedlings was significantly higher than in the distal side. Gene expression of *PsPINs* but not *PsAUX1* is significantly higher than that in the distal one. TIBA did not affect gene expression of *PsPIN1*, *PsPIN2* and *PsAUX1*in the proximal and the distal sides of epicotyls to cotyledon. In addition, altered localization of PsPINs proteins in plasma membrane was observed in hook region but not in middle region of epicotyls in etiolated Alaska pea seedlings, suggesting the lateral auxin distribution from the distal to the proximal sides in epicotyls and resulted in normal graviresponse of etiolated Alaska pea seedlings.

INTRODUCTION

Terrestrial plants, which have evolved under a 1 g gravitational force, have acquired the ability to use and/or to resist gravistimuli to regulate their growth and development. Under the 1 g conditions on Earth, stems and roots of plants grow upwards against, and downwards with, respectively, the direction of gravity due to gravitropic responses enabled by differential cell elongation in these organs. A step forward in the understanding of the cellular mechanisms involved in gravimorphogenesis has already been achieved using microgravity conditions in space, indicating that the morphology of plants is substantially influenced by gravistimulation (Halstead and Dutcher 1987; Brown et al. 1990; Musgrave et al. 1997, 2000; Kiss et al. 1998; Hoson et al. 1999; Takahashi et al. 1999; Kiss 2000). In STS-95 space experiments, we also demonstrated that microgravity conditions substantially affected the growth and development of etiolated pea seedlings (Ueda et al. 1999, 2000). Such morphogenesis observed under microgravity conditions has been designated as automorphogenesis (Hoson et al. 1992; Stankovic et al. 1998). Automorphogenesis-like growth and development of etiolated pea seedlings was demonstrated to be induced when pea plants are constantly rotated three-dimensionally on a three-dimensional (3D) clinostat (Shimazu et al. 2001; Miyamoto et al. 2005). As well as application of the 3D clinostat to simulate microgravity conditions, mutants showing automorphogenesis-like growth and development are valuable in understanding how gravity regulates morphogenesis in plants. One such agravitropic pea mutant, the *ageotropum* pea (from *Pisum sativum* L. cv. Weibull's Weitor), whose roots and shoots lack the ability to orient with respect to gravity (Blixt et al. 1958; Takahashi and Suge 1991), showed automorphogenesis-like epicotyl bending (Hoshino et al. 2007). A possible role for auxin in tropic bending has been proposed by the Cholodny-Went hypothesis, which contends that unequal distribution of auxin between the opposite sides of an organ causes differential cell elongation (Went 1974). Recently, we demonstrated a characteristic asymmetrical polar auxin movement that is gravity-controlled in the early growth stages of etiolated pea epicotyls, and suggested its importance for inducing asymmetrical accumulation of auxin during the negative gravitropic response of the epicotyls, based on the results of transport experiments using radiolabeled auxin (Hoshino et al. 2005). In the present study, to clarify the hypothesis that auxin transport is essential for epicotyl bending for determination of growth direction by gravistimulation in the early growth stage of etiolated pea seedlings, relationships among polar auxin transport in epicotyls, gene expression and the distribution of its product closely related to auxin polar transport were examined.

MATERIALS AND METHODS

Plant materials. Pea (*Pisum sativum* L. cv. Alaska) seeds were set in dry rockwool (Nippon Rockwool Co., Tokyo, Japan) in an acrylic chamber mimicking the plant growth chamber in the STS-95 space experiment (Ueda et al. 1999, 2000), then allowed to germinate and grow after watering with 180 ml of distilled water, as described previously (Ueda et al. 1999, 2000). To study a negative gravitropic response of epicotyls in the early growth stage, the axis of the embryo in dry seed was set horizontally to the rockwool surface. The acrylic chamber in a Zip-lock bag was kept at 23.5°C in the dark. Etiolated pea epicotyls were divided longitudinally into proximal and distal halves after appropriate incubation, frozen in liquid nitrogen and stored at -80°C before use for Northern blot analysis.

Treatment with inhibitors of auxin. Pea seeds were set in a horizontal or inclined position in dry rockwool used for seedling growth in an acrylic chamber, watered with 180 ml of 2,3,5-triiodobenzoic acid (TIBA, Sigma-Aldrich, St. Louis, MO), *N*-(1-

naphthyl)phthalamic acid (NPA, Tokyo Kasei Kogyo, Tokyo, Japan) and 9-hydroxyfluorene-9-carboxylic acid (HFCA, Tokyo Kasei Kogyo, Tokyo, Japan) and *p*-chlorophenoxyisobutylic acid (PCIB, Sigma-Aldrich, St. Louis, MO) at a concentration of 10 μM and allowed to germinate and grow for 48 and 60 h in the dark. The growth direction of epicotyls was determined, and is expressed as the angle between the line of normal morphogenesis and the orientation of the organs in that direction. Each experiment was carried out using at least 10 seedlings. The experiment was repeated three times.

Measurement of polar auxin transport. Measurement of polar auxin transport was performed according to the method reported previously (Hoshino et al. 2006) with some modifications. Segments (10 mm long) of the 1st internodes prepared from 60-h-old Alaska pea seedlings grown under 1 g conditions were prepared. Agar medium (0.9% w/v, 20 μl) containing 1.75 μM (1 μCi mL^{-1}) 3-indoleacetic acid [1-^{14}C] ([1-^{14}C]IAA) (American Radiolabeled Chemical, St. Louis, MO) in Eppendorf tube was applied to the apical (inverted position) or basal (normal position) side of the segments. The tubes were incubated in the dark for 6 h at room temperature. At the end of incubation, a 2-mm piece of the opposite side from the donor side was excised, and the radioactivity was measured directly using a liquid scintillation counter (2200CA, Packard Instrument, Meriden, CT). Almost all the radioactivity in the opposite side of the segments donated radiolabeled IAA in the segments resulted from that of [1-^{14}C]IAA transported within at least 16 h in planta (Oka et al. 1995; Shimazu et al. 2000). Each experiment was carried out using at least ten segments and was repeated three times.

Northern blot analysis. Total RNA was isolated from etiolated pea epicotyls divided into proximal and distal halves of epicotyls to the cotyledons using Isogen (Nippon Gene, Tokyo, Japan) according to the manufacturer's instructions with a minor modification as reported previously (Hoshino et al. 2005). Total RNA was size-fractionated on denaturing 1.0% agarose-formaldehyde gels. Bands of total RNA were transferred onto a nylon membrane (Hybond-N^{+} Amersham Biosciences, Piscataway, NJ) and fixed by baking at 80°C for 2 h after UV cross-linking. Hybridization was achieved with DIG-labeled DNA probes at 58°C using Perfecthyb Plus hybridization buffer (Sigma-Aldrich, St. Louis, MO). DNA probes of PsPINs, PsAUX1 and PsIAA4/5 were prepared as reported previously (Hoshino et al. 2005, 2006). Hybridization membranes were washed with 2 x SSC and 0.1% SDS for 5 min at room temperature, twice with 0.5 x SSC and 0.1% SDS for 20 min at 58°C, and with 0.1 x SSC and 0.1% SDS for 20 min at room temperature. Blots were developed with anti-DIGAP monoclonal antibody (Roche Diagnostics, Penzberg, Germany) with CSPD as the chemilluminescent substrate (Roche Diagnostics) according to the manufacturer's recommendations, and were exposed to X-ray film (RXU, Fuji Photo Film, Tokyo, Japan). Density of blots was quantified using a CS Analyzer ver 2.0 system (Atto, Tokyo, Japan). The amount of ribosomal RNA was used as a loading control. Each experiment used at least ten segments and was repeated three times. Statistical analyses were carried out using Student's t-test for equal variance.

Generation of Antibody and determination of immunolocalization. Since ZmPIN1a (Carraro et al. 2006) showed high homology to PsPINs, ZmPIN1a-specific antibody was decided to introduce for the detection of PsPINs in etiolated Alaska pea seedlings. To generate ZmPIN1a-specific polyclonal antibodies, an oligo-peptide fragment with hydrophilic region was synthesized. The peptide was combined with KLH (keyhole limpet hemocyanin) for the antigen. After immunization of rabbits, the polyclonal antiserum was affinity purified against the ZmPIN1a-specific peptide.

Pea epicotyls were excised from the seedlings and immediately fixed in Carnoy fixative solution (60% Ethanol, 30% Chloroform, 10% Acetic Acid) for 3 h at 4°C maintaining their orientation to gravity. Then samples were extracted from epicotyls and further fixed 90 min. After dehydration through an ethanol and tert-butanol series, the samples were embedded in Paraplast Plus (Sigma-Aldrich). Sections (8–10 μm) were cut using a microtome (RM2135; Leica) and collected on slides. The slides were deparaffinized and treated with detergent solution (10 % DMSO and 3 % Nonidet P-40 in 1-fold MTSB [100 mM piperazine-1,4-bis(2-ethanesulfonic acid) (PIPES) pH 7.2, 10 mM MgSO4 , 10 mM EGTA] for) for 30 min and washed twice in 0.5-fold MTSB for 10 min. Then, the slides were treated with blocking buffer (1-fold MTSB with 1.5 % skim milk) for 30 min and incubated overnight at 4°C with an anti-ZmPIN1a antibody at a 150-fold dilution in blocking buffer. The slides were rinsed three times for 10 min with 0.5-fold MTSB containing 0.1 % Tween-20, and were then incubated with Alexa488-conjugated goat anti-rabbit IgG (Molecular Probes) at a 300-fold dilution for 3 h at room temperature. After washing three times for 10 min, the sample sections were sealed with ProLong Gold Anti slow fade reagent (Molecular Probes). The prepared samples were observed with an epi-fluorescence microscope (model BX51; Olympus, Tokyo, Japan).

RESULTS AND DISCUSSION

Automorphogenesis of etiolated Alaska pea seedlings together with reduced auxin polar transport in epicotyls was observed not only in true microgravity conditions in space (Ueda et al. 1999, 2000) but also in 3D clinostat conditions (Miyamoto et al. 2005) and an agavitropic mutant of ageotropum pea seedlings (Hoshino et al. 2007). Significantly higher auxin polar transport in the proximal side of the 1st internodes to the cotyledons compared to that in the distal side, but not in the 2nd internodes, in etiolated Alaska pea seedlings has also been reported (Hoshino et al. 2006). To clarify close relationships between gravimorphogenesis and auxin polar transport, the effects of inhibitors of auxin polar transport on auxin polar transport, the expression of PsPINs and PsAUX1genes (Chawla and DeMason 2004; Hoshino et al. 2005) related to auxin polar transport in etiolated Alaska pea seedlings were investigated. Together with the expression of PsPINs genes, the localization of PsPIN1 regulating auxin polar transport in epicotyls of etiolated Alaska pea seedlings was also determined. As shown in Table 1, a potent inhibitor of auxin polar transport, TIBA, substantially induced automorphogenesis-like growth and development in etiolated Alaska pea seedlings. Similar powerful inhibitors of auxin polar transport, NPA and HFCA, were found to phenocopy automorphogenesis-like epicotyl bending in etiolated Alaska pea seedlings. However, an inhibitor of auxin action, PCIB, had little effect.

Gene expression detected by Northern blot analysis in the proximal and the distal sides of the 1st internodes of etiolated Alaska pea epicotyls grown on 1 g conditions was investigated. As shown in Fig. 1, gene expression of PsPINs but not PsAUX1 is significantly higher than that in the distal one as well as significant higher auxin polar transport in the proximal side of the 1st internode in etiolated Alaska pea epicotyls (Hoshino et al. 2006).

Fig. 2 shows the effect of TIBA on IAA transport and endogenous levels of auxin detected by PsIAA4/5 gene expression in the 1st internode of etiolated Alaska pea epicotyls. TIBA exogenously applied substantially inhibited auxin polar transport and endogenous levels of IAA detected by gene expression of PsIAA4/5 which is an auxin inducible gene in its concentration dependent manner. Surprisingly TIBA did not affect gene expression of PsPIN1, PsPIN2 and PsAUX1in the proximal and the distal sides of epicotyls to cotyledon although gene expression of PsPIN1 in the proximal side of epicotyls was higher than that in the distal one (Fig. 3).

Localization of PsPINs was immunohistchemically determined with ZmPIN1a antibody, indicating that altered localization of PsPINs indicated by green fluorescence was observed in the proximal and the distal side of epicotyl cells in hook region but not in those of the middle part of the 1st internode in etiolated Alaska pea seedlings (data not shown).

These results described above strongly suggest that auxin polar transport is essential factor for graviresponse of etiolated Alaska pea seedlings. This is also supported by the fact that an accumulation of PsPIN1 mRNA encoding a putative auxin efflux facilitator, which was observed in vascular tissue, cortex and epidermis in the proximal and distal sides of etiolated epicotyls, was markedly influenced by gravistimulation (Hoshino et al 2006). Judging from the fact described above together with the result in this study that altered localization of PsPINs proteins in plasma membrane was observed in hook region but not in middle region of epicotyls in etiolated Alaska pea seedlings, the lateral auxin distribution from the distal to the proximal sides in epicotyls is required for normal graviresponse of etiolated Alaska pea seedlings. A step forward in the understanding of cellular and molecular mechanisms involved in gravimorphogenesis of plants will be achieved in space. The findings of this study substantially lead us to clarify close relationships between auxin polar transport and gravimorphogenesis of plants in cellular and molecular levels using artificial 1 g and true microgravity conditions on the Japanese Experiment Module "Kibo" in the International Space Station in near future.

ACKNOWLEDGEMENT

This study was partially supported by JSPS KAKENHI Grant numbers 23510260 and 26506018.

LITERATURE CITED

Blixt S, Ehrenberg L, Gelin O (1958) Quantitative studies of induced mutations in peas.

I. Methodological investigation Agri Hortic Genet 16: 238-250.

Brown AH, Chapman DK, Lewis RF, Venditti AL (1990) Circumnutations of sunflower hypocotyls in satellite orbit. Plant Physiol 94: 233-238.

Carraro N, Forestan C, Canova S, Traas J, Varotto, S (2006) ZmPIN1a and ZmPIN1b encode two novel putative candidates for polar auxin transport and plant architecture determination of maize. Plant Physiol 142: 254-264.

Chawla R, DeMason DA (2004) Molecular expression of PsPIN1, a putative auxin efflux carrier gene from pea (Pisum sativum L.). Plant Growth Regul 44: 1-14.

Halstead TW, Dutcher FR (1987) Plants in space. Annu Rev Plant Physiol 38: 317-345.

Hoshino T, Hitotsubashi R, Miyamoto K, Tanimoto E, Ueda J (2005) Isolation of PsPIN2 and PsAUX1 from etiolated pea epicotyls and their expression on a three-dimensional clinostat. Adv Space Res 36: 1284-1291.

Hoshino T, Miyamoto K, Ueda J (2006) Requirement for the gravitycontrolled transport of auxin for a negative gravitropic response of epicotyls in the early growth stage of etiolated pea seedlings. Plant Cell Physiol 47: 1496-1508.

Hoshino T, Miyamoto K, Ueda J. (2007) Gravity-controlled asymmetrical transport of auxin regulates a gravitropic response in the early growth stage of etiolated pea (Pisum sativum) epicotyls: studies using simulated microgravity conditions on a three-dimensional clinostat and using an agravitropic mutant, ageotropum. J Plant Res 120: 619-628.

Hoson T, Kamisaka S, Masuda Y, Yamashita M (1992) Changes in plant growth processes under microgravity conditions simulated by a three-dimensional clinostat. Bot Mag Tokyo 105: 53-70.

Hoson T, Soga K, Mori R, Saiki M, Wakabayashi K, Kamisaka S, Kamigaichi S, Aizawa S, Yoshizaki I, Mukai C, Shimazu T, Fukui K, Yamashita M (1999) Morphogenesis of rice and Arabidopsis seedlings in space. J Plant Res 112: 477-486

Kiss JZ (2000) Mechanisms of the early phases of plant gravitropism. CRC Crit Rev Plant Sci 19: 551-573.

Kiss JZ, Katembe WJ, Edelmann RE (1998) Gravitropism and development of wild-type and starch-deficient mutants of Arabidopsis during spaceflight. Physiol Plant 102: 493-502.

Miyamoto K, Hoshino T, Yamashita M, Ueda J (2005) Automorphosis of etiolated pea seedlings in space is simulated by a threedimensional clinostat and the application of inhibitors of auxin polar transport. Physiol Plant 123: 467-474.

Musgrave ME, Kuang A, Matthews SW (1997) Plant reproduction during spaceflight: importance of the gaseous environment. Planta 203: S177-S184.

Oka M, Ueda J, Miyamoto K, Yamamoto R, Hoson T, Kamisaka S. (1995) Effect of simulated microgravity on auxin polar transport in inflorescence axis of Arabidopsis thaliana. Biol Sci Space 9: 331-336.

Shimazu T, Miyamoto K, Hoson T, Kamisaka S, Ueda J (2000) Suitable experimental design for determination of auxin polar transport in space using a spacecraft. Biol Sci Space 14: 9-13.

Shimazu T, Yuda T, Miyamoto K, Yamashita M, Ueda J (2001) Growth and development in higher plants under simulated microgravity conditions on a 3-dimensional clinostat. Adv Space Res 27: 995-1000.

Stankovic B, Volkmann D, Sack FD (1998a) Autotropism, automorphogenesis, and gravity. Physiol Plant 102: 328-335.

Takahashi H, Mizuno H, Kamada M, Fujii N, Higashitani A, Kamigaichi S, Aizawa S, Mukai C, Shimazu T, Fukui K, Yamashita M (1999) A spaceflight experiment for the study of gravimorphogenesis and hydrotropism in cucumber seedlings. J Plant Res 112: 497-505.

Takahashi H, Suge H (1991) Root hydrotropism of an agravitropic pea mutant, ageotropum. Physiol Plant 82: 24-31.

Ueda J, Miyamoto K, Yuda T, Hoshino T, Fujii S, Mukai C, Kamigaichi S, Aizawa S, Yoshizaki I, Shimazu T, Fukui K (1999) Growth and development, and auxin polar transport in higher plants under microgravity conditions in space: BRICAUX on STS-95 space experiment. J Plant Res 112: 487-492.

Ueda J, Miyamoto K, Yuda T, Hoshino T, Sato K, Fujii S, Kamigaichi S, Izumi R, Ishioka N, Aizawa S, Yoshizaki I, Shimazu T, Fukui K (2000) STS-95 space experiment for plant growth and development, and auxin polar transport. Biol Sci Space 14: 47-57.

Went FW (1974) Reflections and speculations. Annu Rev Plant Physiol 25: 1-26.

Table 1. Effect of inhibitors of auxin polar transport or auxin action, 10 μM of TIBA (2,3,5-triiodobenzoic acid), NPA (*N*-(1-naphthyl)phthalamic acid), HFCA (9-hydroxyfluorene-9-carboxylic acid), and PCIB (*p*-chlorophenoxyisobutylic acid), respectively, on gravimorphogenesis of epicotyls in etiolated Alaska pea seedlings.

Treatment	Angle, degree[z]
Control	12.8±1.4
TIBA	43.5±1.7*[y]
NPA	55.6±3.1*
HFCA	49.1±2.1*
PCIB	13.7±1.3

[z] The growth direction of epicotyls was expressed as the angle between a vertical line and the orientation of the organs in the direction. Values are the mean of at least 20 seedlings with standard errors.

[y] Statistical analyses were carried out using Student's *t*-test ($P < 0.01$). Values marked with an asterisk are significantly higher or lower than DMSO treatment control.

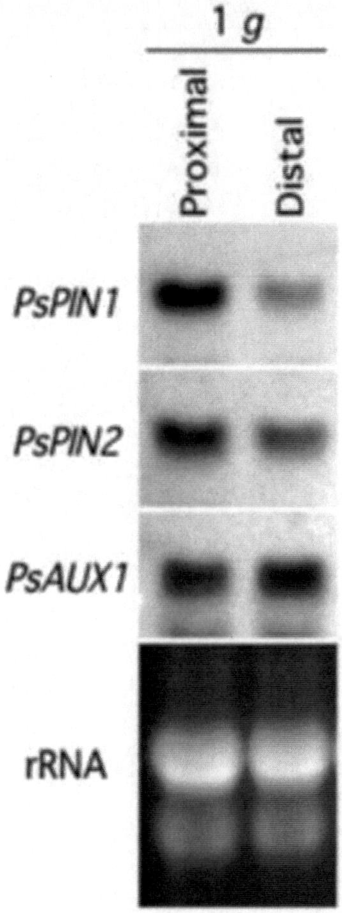

Figure 1. Gene expression of *PsPINs* and *PsAUX1* in the 1st internode of etiolated Alaska pea epicotyls detected by Northern blot analysis.

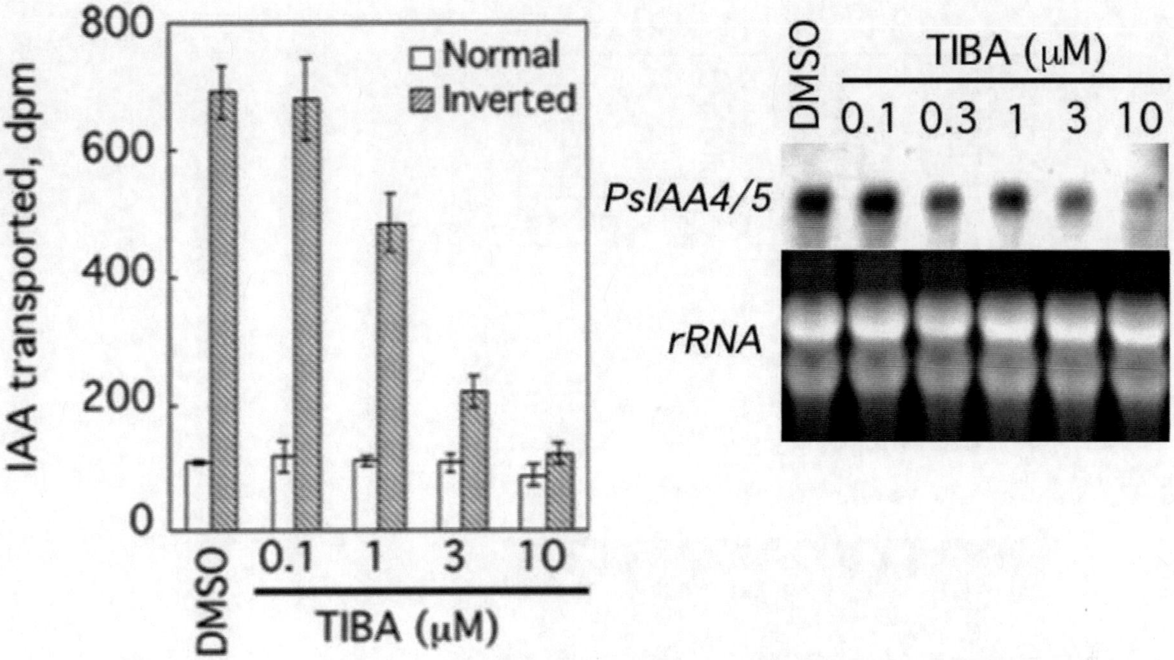

Figure 2. Effect of TIBA on auxin polar transport (left) and endogenous levels of IAA detected by gene expression of *PsIAA4/5* (right) in etiolated Alaska pea epicotyls. Radiolabeled IAA was applied to the apical (inverted) or basal (normal) side of the segments. Results are expressed as averages of three independent experiments with standard errors of the mean.

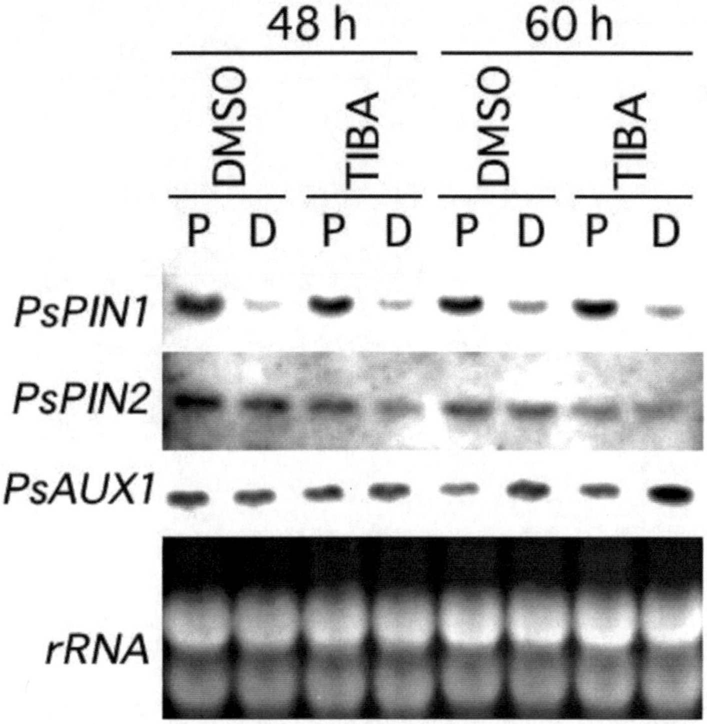

Figure 3. Effect of TIBA on gene expression of *PsPINs* and *PsAUX1* detected by Northern blot analysis in etiolated Alaska pea epicotyls. Segments (10 mm long) of the 1st internodes excised from 48 or 60-h-old Alaska pea seedlings grown under 1 g conditions were divided into proximal (P) and distal (D) halves of epicotyls to cotyledons.

HELIOLACTONE, A NON-SESQUITERPENE LACTONE GERMINATION STIMULANT FOR ROOT PARASITIC WEEDS FROM SUNFLOWER

Kotomi Ueno[1,4], Shuhei Umeda[1], Masaharu Mizutani[1], Hirosato Takikawa[1], Yukihiro Sugimoto[1], Toshio Furumoto[2] and Rossitza Batchvarova[3]

[1]Kobe University, Nada, Kobe, Japan

[2]Kagawa University, Miki, Kagawa, Japan

[3]AgroBio Institute, 1164 Sofia, Bulgaria

[4]Corresponding author: ueno@crystal.kobe-u.ac.jp

Root exudates of sunflower (*Helianthus annuus* L.) line 2607A exhibited germination inducing activity on the seeds of root parasitic weeds *Orobanche cumana*, *O. minor*, *O. crenata*, *Phelipanche aegyptiaca,* and *Striga hermonthica*. Bioassay-guided purification led to the isolation of a germination stimulant designated as heliolactone. FT-MS analysis indicated that the molecular formula of heliolactone is $C_{20}H_{24}O_6$. Detailed NMR studies revealed that heliolactone has a methylfuranone group, a common structural component of strigolactones, which is connected to a C_{15} part via an enol ether bridge. Heliolactone induced seed germination of the above mentioned root parasitic weeds, while dehydrocostus lactone and costunolide, sesquiterpene lactones isolated from sunflower root exudates, were effective only on *O. cumana* and *O. minor*. Heliolactone production in aquacultures increased when sunflower plants were grown hydroponically in tap water. Supplementation with either phosphorus or nitrogen reduced the stimulant production. Costunolide was detected at a higher concentration in well-nourished medium as opposed to nutrition-deficient media, suggesting the contrasting contribution of heliolactone and the sesquiterpene lactone to the germination induction of *O. cumana* under different soil fertility levels.

EXPEDITED IN VITRO BALANCED SEEDLING PRODUCTION OF ALEXANDRIAN LAUREL USING BA AND TDZ

Guochen Yang[1,2] and Zhongge (Cindy) Lu[1]

[1]North Carolina A&T State University, Greensboro, NC, USA

[2]Corresponding author: yangg@ncat.edu

An efficient micropropagation protocol was developed to rapidly produce Alexandrian laurel (*Danae racemosa* L.), a commercially popular dark evergreen shrub. Our micro-propagated plantlets also produce more vigorous growth characterized by more substantial roots than conventionally seed-propagated plants. We have successfully improved the growth process using BA and TDZ, to increase shoot multiplication and enhance seedling quality characterized by a good balance of shoot and root growth. The effect of introducing BA has been to balance seedling development by simultaneously accelerating shoot growth and slowing down root growth, whereas the inclusion of TDZ has promoted shoot multiplication and proliferation by producing more than 40 shoots per seed. The micro-propagated plantlets developed using this protocol have been successfully acclimatized and established outside the greenhouse.

THE APPLICATION OF PLANT DEFENSE ELICITORS AND BIOPESTICIDES FOR MANAGING DISEASE AND IMPROVING CROP YIELD AND QUALITY

Gary E. Vallad[1,2]

[1]*University of Florida Gulf Coast Research and Education Center, Wimauma, FL, USA*

[2]*Corresponding author: gvallad@ufl.edu*

Several plant defense elicitors and numerous natural products (biopesticides) are commercially available for managing diseases of numerous crops. Although these products offer the promise to offset or augment the use of conventional pesticides, they are not without their specific challenges. Information regarding the relative efficacy or 'best usage pattern' for many products remains limited. This is especially true for biopesticides where the actual mode(s) of action remain unknown or presumed based on similarity to other characterized compounds and microorganisms; as such, the recommended usage-patterns may be incompatible with their true mode(s) of action. Additional research is also needed to determine the compatibility of biopesticides with their environment and potentially with the plant host. Defense elicitors have also shown success for managing a variety of plant diseases, but must be applied carefully to avoid unintended impacts on plant yield. The public's demand for low-risk and environmentally-friendly pesticides must be balanced with the grower's need for products that are cost-effective; especially as global markets continue to expand and compete for limited resources. Therefore, the need for additional research is sure to continue, not only to address the above-mentioned challenges, but to develop the next generation of defense elicitors and biopesticides with improved efficacy.

RyzUp SmartGrass® Effects on Smooth Brome Growth and Yields

Michael D. Rethwisch[1,2], Garrett Tooker[1], Scott Willet[1] and Craig Hruska[1]

[1]University of Nebraska- Lincoln, Butler County Extension, 451 N. 5th Street, David City, NE 68632

[2]Corresponding author: mrethwisch2@unl.edu

Various aspects of smooth brome (*Bromus inermis*) growth were documented in response to various RyzUp SmartGrass® application aspects (differing rates, surfactants, soil fertility) during 2013 and 2014 in southeastern Nebraska. Application of RyzUp SmartGrass® treatments resulted in highly visible and significantly taller forage heights for 30 days post treatment, while chlorophyll concentration levels were reduced. RyzUp SmartGrass® treatments increased hay production by up to 49% at 30 days post treatment. Application to highly fertilized hay fields resulted in 7-14% more hay at 60 days post treatment, with less response noted at lesser fertility levels. Treatments also resulted in numerically more stems/area in highly fertilized areas, but not other sites. Differences in growth responses were noted from various surfactants when used with RyzUp SmartGrass® and at approximately 30 days post treatment, however, significantly greater brome production was noted when ClassAct NG was used with RyzUp SmartGrass®.

INTRODUCTION

Recent high forage prices resulted in increased interest and need to economically increase forage production, especially for that of various grass hay species such as smooth brome (*Bromus inermis*). Initial screening research had documented that RyzUp SmartGrass® (active ingredient = gibberellic acid 3, Valent USA) was highly effective in increasing smooth brome production, however, additional data were needed for smooth brome growth response to other aspects (rate, surfactants, etc.) of RyzUp SmartGrass®.

This project evaluated smooth brome growth in response to application of three rates of RyzUp SmartGrass® and the effect of various surfactants when applied to smooth brome hay fields that had varying levels of soil fertility.

METHODS AND MATERIALS

Three rates (0.3, 0.6 and 0.9) oz./acre of RyzUp SmartGrass® were applied to smooth brome in four locations in eastern Nebraska in the early spring of 2013 and 2014. Locations varied in fertility levels and plant growth stage, with one site (Rising City) being almost deficient in soil nitrates and phosphorus levels at time of application, one site (Dorchester) had medium levels of soil nitrates, while both David City locations were well fertilized. The Rising City location was fertilized with a high rate of ammonium sulfate approximately 5 days after RyzUp SmartGrass® application to provide necessary nitrogen, however, phosphorus was not included for plot fertilization at this site.

Three differing surfactants (Kinetic, Helena Company, Memphis, TN; BioLink Spreader Sticker, Westbridge Agricultural Products, Vista, CA; AgriSolutions ClassAct NG, Winfield

Solutions, LLC, St. Paul, MN) were also evaluated with the low rate of RyzUp SmartGrass® at the three 2014 locations (David City, Dorchester, Rising City) as had been done at a 2013 location (Brainard).

Kinetic consists of 99% polyethylenepolyoxypropylene copolymer, polysiloxane polyether copolymer. It was used at the rate of 6 oz./100 gallons.

ClassAct NG consists of 50.5% ammonium sulfate, corn syrup and alkyl polyglycoside. It was used at 2.5% v/v, and provided 0.5 lbs./acre of nitrogen and 0.57 lbs./acre of sulfur.

BioLink spreader-sticker was used at 4 oz./acre (1.12 oz./acre), and consists of 10.1% soapbark in addition to alkylphenol ethoxylate and polysaccharide.

Treatments were applied with a backpack sprayer calibrated to deliver 28 gallons/acre to plots that were 7 foot wide x 25 foot long. Each treatment had four replications at each site utilizing a randomized complete block design.

Plant data were collected weekly for standing forage heights and extended leaf heights for approximately 30 days after application. Forage yields were obtained at approximately 30 and 60 days post treatment in 2013.

Results

Rate responses were noted in both 2013 and 2014, with taller forage height and extended leaf heights trending higher as rate increased early after application. This was noted at all sites (Figs. 1-3).

An early growth response trend was also noted among the different surfactants, with greatest early growth noted when ClassAct NG was utilized (Fig. 2-4).

Differences were visibly noted for approximately 21 days after treatment, with differences readily available at some locations beyond this.

The visible differences were also noted in smooth brome hay yields obtained from the 2013 experimental site (Brainard) where the surfactants were evaluated. Highest mean smooth brome yields at 28 days post treatment among the surfactants used with RyzUp SmartGrass® was noted from usage of ClassAct NG (2,995 lbs./acre), followed by usage of BioLink and then Kinetic (Fig. 5). Untreated smooth brome had significantly less yield (2,363 lbs/acre), while addition of 20-20-20 fertilizer also provided a growth response on this site which was slightly deficient for soil phosphorus.

Summary

Smooth brome growth was shown to responsive positively as rate of RyzUp SmartGrass® increased at multiple locations during this study.

Yield data documented an 11% increase from usage of all rates of RyzUp SmartGrass® on smooth brome in 2013 at the David City location, however this amount of increase was not noted in hay fields with less than optimum fertility.

Data also indicated that utilizing differing surfactants with this product also resulted in differences in early smooth brome growth. This was noted in all four experiments which evaluated surfactants while applying RyzUp SmartGrass®.

The reason for differences in smooth brome growth is unknown. The surfactant in these experiments which resulted in most growth was ClassAct NG. This product does contain nitrogen. It was noted that the addition of 20-20-20 to RyzUp SmartGrass® also increased growth. Further experimentation is necessary to elucidate the role of nitrogen may have in efficacy of RyzUp SmartGrass®.

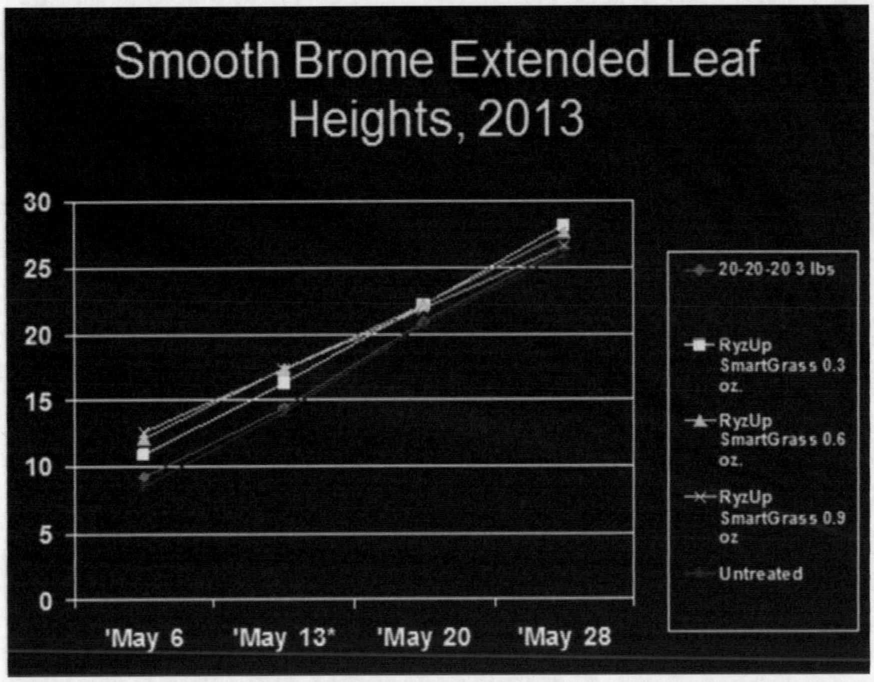

Figure 1. Mean extended leaf height smooth brome heights resulting from application of three rates of RyzUp SmartGrass®, David City, NE, 2013.

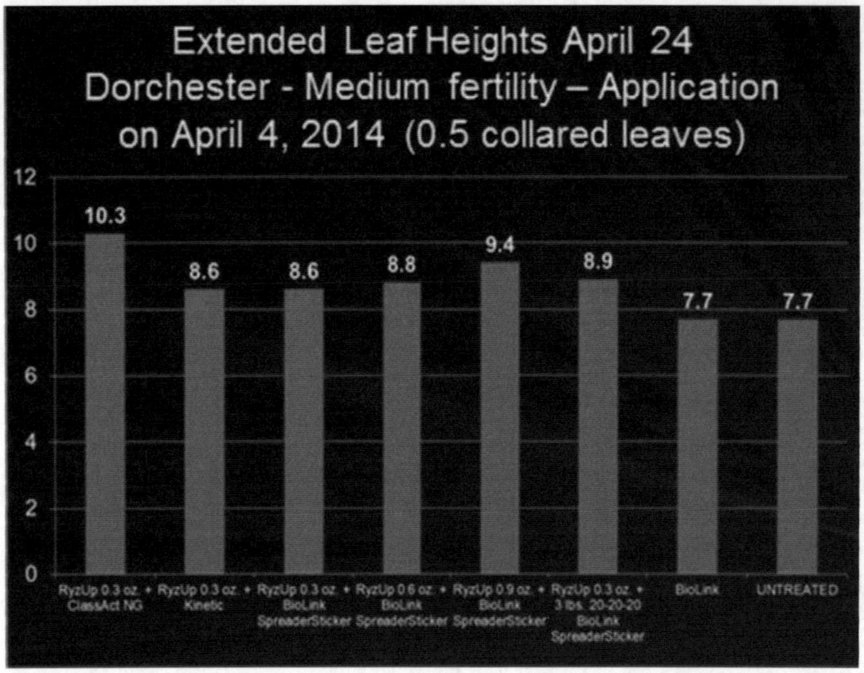

Figure 2. Mean extended leaf heights resulting from application of three rates of RyzUp SmartGrass® and differing surfactants, Dorchester, NE, 2014.

Figure 3. Mean extended leaf height smooth brome heights resulting from application of three rates of RyzUp SmartGrass® and three surfactants, David City, NE, 2014.

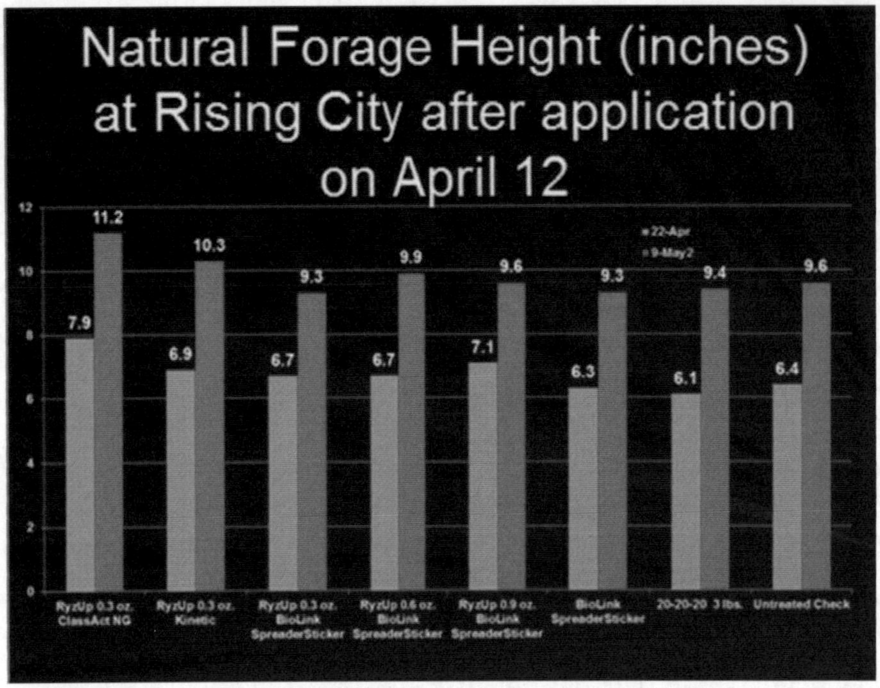

Figure 4. Mean natural forage and extended leaf height smooth brome heights resulting from application of three rates of RyzUp SmartGrass® and three surfactants, Rising City, NE, 2014.

Figure 5. Mean smooth brome hay yields (lbs./acre) at 28 days post treatment with fertilizer and/or RyzUp SmartGrass® on April 24, 2013, Brainard, NE.

APPLICATIONS OF STIMPLEX™, A COMMERCIAL BIOSTIMULANT, IMPROVES RED COLOR ON 'JONAGOLD' APPLES

Holly Little[1,2]

[1]*Acadian Seaplants, 2443 Fair Oaks Blvd. #67, Sacramento, CA 95825 USA*

[2]*Corresponding author:* *hlittle@acadian.ca*

Peel color in apples is an important quality attribute, and is important in determining market acceptance. Well colored, bright red apples are generally preferred by consumers. Red color development is dependent on environmental conditions, and chlorophyll degradation. Anthocyanins, carotenoids, and flavonoids all are pigments involved in coloration. Temperature, light, crop load, water stress, and plant growth regulators can all have an impact on color development.

The use of seaweed in agricultural production has been used since historical times, however its use expanded after a practical extraction method was developed in 1949, which allowed for easier transportation of active compounds in seaweeds to areas further from the coast. The brown seaweed, *Ascophyllum nodosum*, is considered the most researched seaweed for use in agriculture. Stimplex™ (*Ascophyllum nodosum* extract, Acadian Seaplants), is made through a non-pressurized alkaline extraction process, and contains no additives. It has been documented that different species of seaweed and different extraction processes create products with a different chemical makeup that will likely have different activity when applied to plants.

Prohydrojasmon (Blush, Fine Americas, Inc) (PDJ) has recently been registered for increasing red color in apples. PDJ is a synthetically produced jasmonate, which functions as a functional analogue of jasmonic acid in plants. Jasmonates are involved in fruit de-greening by enhancing chlorophyll degradation, as well as enhancing anthocyanin and carotene accumulation.

Both Blush and Stimplex have been used for improving fruit color in apples.

Field trials in Wayne County, NY began in 2012 to evaluate the effect of Stimplex™ (*Ascophyllum nodosum* extract, Acadian Seaplants) on red color development in 'Jonnagold' apples. A full season Stimplex program was compared to a grower standard program, which included an additional biostimulant. The percent red fruit color was significantly higher in fruit from Stimplex treated trees than in the grower standard (Figure 1). In 2013 the same trees received the same treatments, but plots were split, and half the plots received a preharvest treatment of Blush. Without the application of Blush, Stimplex treated trees had a numerically higher percent red color compared to the grower standard (Figure 2). Percent red color was higher in 2013 than 2012. Percent red color with Blush was numerically the same at Stimplex, but significantly greater than the grower standard control (Figure 2). The percent red color in the Stimplex plus Blush treatment was significantly greater than the other treatments.

These results indicate the benefits of a Stimplex program, under commercial production practices, for improving peel color in Jonagold apples with or without the application of Blush. The mechanism of red color improvement by Stimplex in apples has not been determined, however other trials have shown increases in phenolic compounds, fand flavonoids resulting in increased antioxidant activity, as well as increased jasmonic acid under disease pressure.

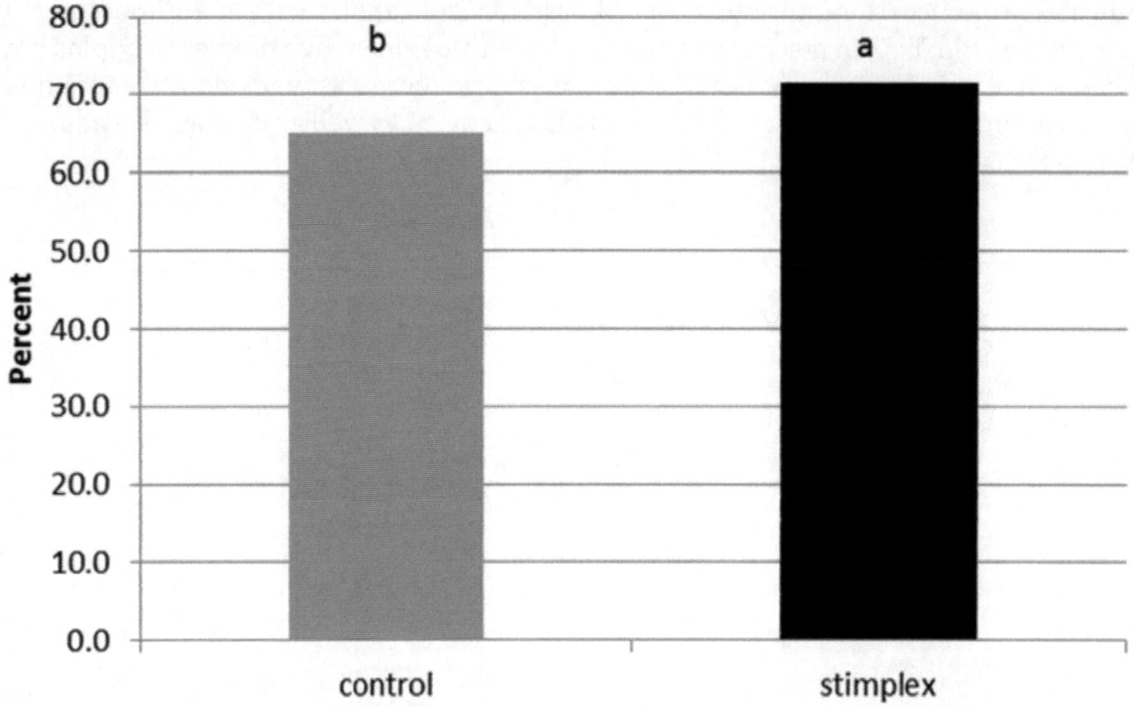

Figure 1. In 2012, the first year of the trial, grower standard was compared to Stimplex program. Percent red color was greater in the Stimplex treatment (p=0.05).

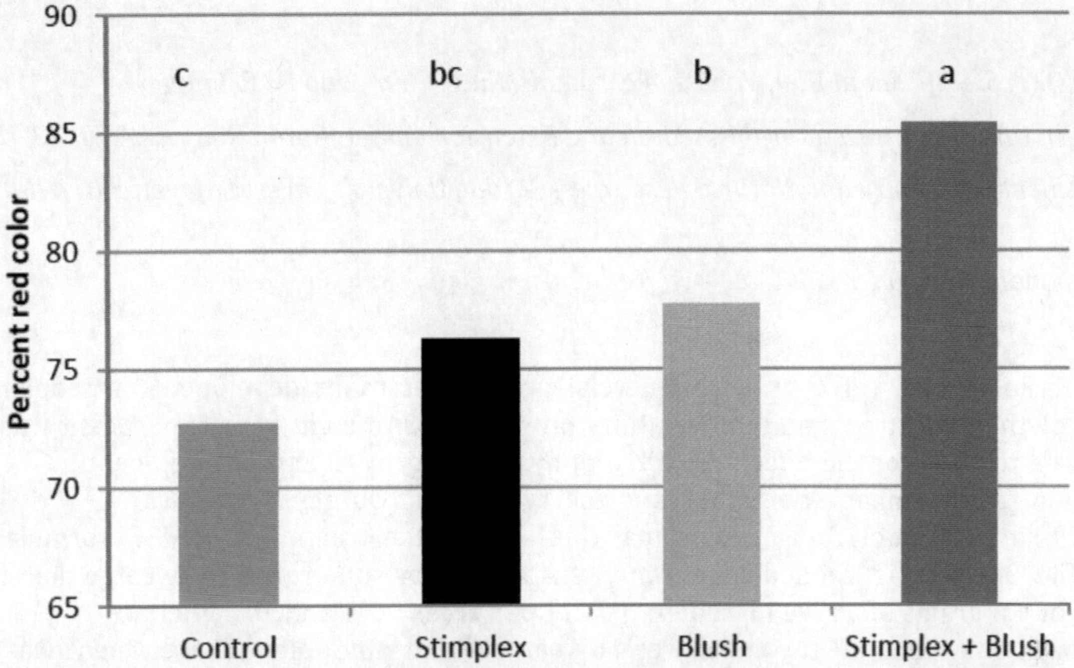

Figure 2. In 2013, an additional treatment of Blush (prohexidine jasmonate) was added. The greatest improvement in fruit color over the control was seen in the combined Stimplex blush treatment. Stimplex alone did not show a difference in fruit color compared to the blush or the grower standard control (p=0.05).

EFFECTS OF PACLOBUTRAZOL TREE GROWTH REGULATOR ON TWO TREE SPECIES

Yadong Qi[1,3], S. Self[1], Shuju Bai[1], Vanessa Ferchaud[1], Kit L. Chin[1], and W. R. Chaney[2]

[1]Southern University Agricultural Research and Extension Center, Baton Rouge, LA 70813, USA

[2]Department of Forestry and Natural Resources, Purdue University, West Lafayette, IN 47907, USA

[3]Corresponding author: yadong_qi@suagcenter.com; yadong_qi@subr.edu

Paclobutrazol (PBZ) is a commercial growth retardant developed for reducing tree growth along streets and under utility power lines. An eight-year experiment was conducted to monitor the effects of PBZ on physiology, growth and development of sweetgum (Liquidambar styraciflua) and cherrybark oak (Quercus falcata var. pagodafoila L.) in Baton Rouge, Louisiana, USA. The initial treatment of PBZ, formulated as Profile 2SC®, was applied in a water suspension by soil drench to sweetgum at a dosage of 4.8 grams of active ingredient (g a.i.) per tree, and 9.6 g a.i. to cherrybark oak when all the trees were 6-year old and no additional treatment was applied thereafter. The PBZ treated trees in both species exhibited smaller leaf size, thicker leaves, shorter leaf petioles, higher chlorophyll content, higher light absorbance, darker green foliage and more compact crowns with improved gas exchange rates than the control trees. But sweetgum trees were more sensitive to PBZ treatment than cherrybark oak trees. The eight year study indicated PBZ application significantly reduced the tree height by 74% and DBH by 76% in sweetgum while only by 34% and 45% in cherrybark oak, respectively. The improved physiological performance and reduced tree size can help enhance tree vigor and at the same time lower the pruning associated maintenance cost. Thus, applied properly, PBZ can be used as an effective tool in urban and community tree management.

INTRODUCTION

Due to the high cost for frequent mechanical trimming of trees under electric power lines, the electric industry funded research in the late 1950s to investigate chemical control of growth following trimming. In the 1970s, tree growth retardants (TGRs) and more economical application techniques were used, and paclobutrazol (PBZ) was among the tree growth retardants developed for use in utility forestry and later extended in arboricultural and urban forestry practices. However, more research is still needed to evaluate the dosage effect of PBZ on growth control of different species of trees over time.

Since 1997, we at Southern University in Baton Rouge, LA in collaboration with scientists at Purdue University in West Lafayette, IN have investigated the effects of PBZ on physiological, biochemical, anatomical and growth performance of different urban tree species in both locations and reported our findings in various publications (Bai, 1999; Bai et al., 1998a, 1998b, 2004, 2005; Chaney, 2001, 2003, 2005; Qi et al., 2000, 2001, 2002). Our research indicated that PBZ treated trees exhibited smaller leaf size, thicker leaves, shorter leaf petioles, higher chlorophyll content, higher light absorbance, darker green foliage and more compact crowns with improved gas exchange rates than untreated control trees (Qi et al., 2000, 2001, 2002), and PBZ treated trees appeared to have enhanced vigor, which is beneficial for trees to be tolerant to heat and drought stresses in urban conditions.

We continued our data collection to understand the long term effects of PBZ on growth performance of different tree species. This paper is part of a follow-up study that reports the findings of an 8-year experiment of the effects of PBZ on height, diameter and cambium growth of two urban tree species, sweetgum (Liquidambar styraciflua) and cherrybark oak (Quercus falcata var. pagodafoila L.).

Material and Methods

This experiment was established in the field of the Southern University Horticulture Farm in Baton Rouge in March 1997 in collaboration with Purdue University. Two tree species, sweetgum and cherrybark oak, with fourteen saplings each were chosen for the experiment. Half of the sapling was randomly selected and treated with PBZ. PBZ, formulated as Profile 2SC®, was applied in a water suspension by soil drench to sweetgum at a dosage of 4.8 grams of active ingredient (g a.i.) per tree and 9.6 g a.i. per tree for cherrybark oak, and these dosages were determined according to the product label. The saplings were all six years old at the time of the treatment and no additional treatment was applied. Tree diameter and height of the trees were measured in the beginning of the experiment in 1997, two years later in 1999, and eight years later in 2005. The diameter was measured at the 10cm trunk level above the ground. The increment borer was used to obtain the cambium growth cores that were imaged using a stereomicroscope (Olympus SZ12).

Results and Discussion

Height Reduction. PBZ was applied to both species in 1997 and tree height was first measured in 1999 (two years later) and again in 2005 (eight years later) (Table 1). The results indicated after 2 years of PBZ treatment, tree height was significantly reduced in both species, with sweetgum reduced by 40% and cherrybark oak by 20%. But the results after 8 years of PBZ treatment somewhat varied with the species. In sweetgum, significant difference still existed between control and treated trees at the end of 8 years treatment with overall height reduction of 74%, while in cherrybark oak, despite an average 34% height reduction in treated trees, no significant difference existed due to large variations occurred in height measurements, indicating the PBZ effect was fading away in some treated cherrybark oak trees after the eight years and re-treatment may be needed (Table 1).

Diameter Reduction. The eight year experiment indicated PBZ treatments significantly reduced the diameter growth in both species (Figures 1 and 2). In sweetgum (Figure 1), two years application of PBZ resulted in 41% reduction in diameter and eight years application resulted in 76% reduction in diameter. In fact, the diameter of treated sweetgum trees only increased by 1.03cm (6.06cm in 2005 minus 5.03cm in 1997) in the eight years, while the control trees increased by 19.74cm (25.04cm in 2005 minus 5.30cm in 1997) during the same period. This indicated the strong cambium growth control effect caused by the PBZ dose (4.8 g a. i. per tree) applied to sweetgum, which, interestingly, was only half of the dose applied to cherrybark oak (9.6 g a.i. per tree). In cherrybark oak (Figure 2), the PBZ application resulted in 45% reduction in diameter over the eight years. So contrasting to cherrybark oak, sweetgum appeared to be more sensitive to PBZ application. This sensitivity may be attributed to the differences in their cambium growth, xylem vessel distribution and water uptake patterns as well as their biochemical pathway in reaction to PBZ. In addition to the strong growth control resulted from application of PBZ in sweetgum, it was not without question. It was observed that multiple branches were developed at the bases of several treated trees in sweetgum that were not desirable and affected the aesthetic appearance of the trees. However, despite the growth reduction, PBZ application did not affect the aesthetic and overall physical appearance of cherrybark oak.

Cambium Growth Reduction. The effect of PBZ on cambium growth reduction is illustrated using cherrybark oak as an example (Figure 3). As shown in Fig. 3 top picture, a non-treated cherrybark oak tree's one year annual increment as result of its cambium growth was almost equivalent to the total cambium growth of 5-6 years combined in a PBZ treated tree (Fig. 3, bottom picture). The PBZ effect was strong during the first 4 years and began to fade away after 5 years (bottom pic). As such, reapplication of PBZ should be considered every 5-6 years for cherrybark oak. While PBZ is at work (Fig. 3, bottom picture), it remarkably compacted the development of vessel members in the early wood and significantly limited the later wood development; therefore, reducing the diameter growth. This wood internal structure provided a fingerprint on how PBZ affected the cambium development and thus reduced diameter growth of a tree. It further illustrated that PBZ reduced cambium growth largely through limiting the late wood development in cherrybark oak, while early wood still possessed prominent numbers of ring porous vessels, which facilitated water uptake effectively. Thus the reduction in cambium growth probably did not negatively affect water uptake. Rather PBZ application may have enhanced the water uptake because of more vessel members per unit of xylem area appeared in the treated trees, and this was witnessed through the significantly enhanced transpiration rate observed in PBZ treated trees (Qi et al., 2002).

Conclusions

The eight year study indicated PBZ application significantly reduced the tree height by 74% and diameter by 76% in sweetgum. Sweetgum appeared to be very sensitive to PBZ application. PBZ dosage at 4.8.g a.i used in this study may be too high and should be reduced. A further study is needed to develop suitable application rate for sweetgum. The effect of cherrybark oak modified by PBZ at 9.6g a.i. was considered moderate, showing 45% reduction in diameter but somewhat insignificant height reduction at the end of eight years. The anatomy of the cambium growth in cherrybark oak indicated the effect of the PBZ application at 9.6 g a.i. lasted approximately 5 years. Reapplication of PBZ may be considered for usage every 5-6 years. Of course, the best rate should always be evaluated based on tree species, age, and climate conditions. In general, PBZ and other TGRs should be systematically evaluated against different tree or woody species in order to be widely used as an effective tool for utility and urban forestry management.

Acknowledgements

This research was partially sponsored by the grants from USDA/NIFA McIntire-Stennis Formula Fund to Southern University Agricultural Research and Extension Center, Purdue University, and Southern University NSF funded SMART Program.

Literature Cited

Bai, S., Chaney, W. R., Y. Qi, and Harvey A. Holt. 1998. Effects of paclobutrazol on Closure of Wounds in Trees. Proceedings of Western Plant Growth Regulator Society, Vol. 10:12-18.

Bai, S., W. R. Chaney, and Y. Qi. 1998. Wound Closure in Trees Affected by Paclobutrazol Growth Regulator. pp. 370-371. Proceedings of Society of American Foresters 1997 National Conference, Memphis, Tennessee, October 4-8, 1997.

Bai, Shuju. 1999. Effects of tree growth regulators on several tree species. Ph.D. Dissertation, Purdue University, West Lafayette, IN.

Bai, S. W.R. Chaney, and Y. Qi. 1999. Closure of bark wounds affected by paclobutrazol, pp. 227-233. IN: Proceedings of Plant Growth Regulation Society of America 26th Annual Meeting, Costa Mesa, CA July 11-14, 1999.

Bai, S., W. R. Chaney, Y. Qi. 2004. Response of cambia and shot growth in trees treated with paclobutrazol. Journal of Arboriculture and Urban Forestry 30: 137-145.

Bai, S., W. R. Chaney, Y. Qi. 2005. Wound closure in trees affected by paclobutrazol. Journal of Arboriculture and Urban Forestry 31:273-279

Chaney, W. R. 2003. Tree growth retardants: Arborists discovering new uses for an old tool. Tree Care Industry. p54-58.

Chaney, W. R. 2005. A paclobutrazol treatment can leave a tree more stress tolerant. Golfdom. February, 2005. Publication by Advanstar Communications Inc.

Chaney, William R. 2001. Tree growth retardants provide multiple benefits. Under the Canopy. p9-11. http://uaf.edu/files/ces/newsletters/underthecanopy/0107canopy.pdf

Qi, Y., S. Bai, W. R. Chaney, K. Drye. 2000. Effect of paclobutrazol on leaf optical properties of sweetgum and cherrybark oak. Proceedings of 27th Annual Meeting of Plant Growth Regulation Society of America (ed. Richard Dunand). pp260-265

Qi, Y., S. Bai, W. R. Chaney, J. Qin. 2001. Effects of paclobutrazal on leaf UV-B absorbing compounds and chlorophyll concentrations and leaf anatomy of sweetgum and cherry bark oak. Proceedings of the 28th Annual Meeting of Plant Growth Regulation Society of America. pp107- 113.

Qi, Yadong, T. Knighten, W. R. Chaney. 2002. Leaf physiology and anatomy of two urban tree species as affected by paclobutrazol tree growth regulator. Proceeding of the 29th Annual Meeting of Plant Growth Regulation Society of America. p123-128.

Table 1. Height measurements (cm) in 1999 and 2005 and total percentage of reduction.

Treatment	Height (cm)		Percentage of total reduction
	1999	2005	
Sweetgum			
Control	425.00a	1039.37a	74
Treated	255.00b	273.30b	
Cherrybark Oak			
Control	305.00a	847.34a	34
Treated	245.00b	557.78a	

Values followed by the same lower case letter for each pair are not significantly different at $P \leq 0.05$ level.

Note: The PBZ treatment was initiated in March 1997.
1999 – the measurement was taken at the end of 2 years of the PBZ treatment.
2005 – the measurement was taken at the end of 8 years of the PBZ treatment.

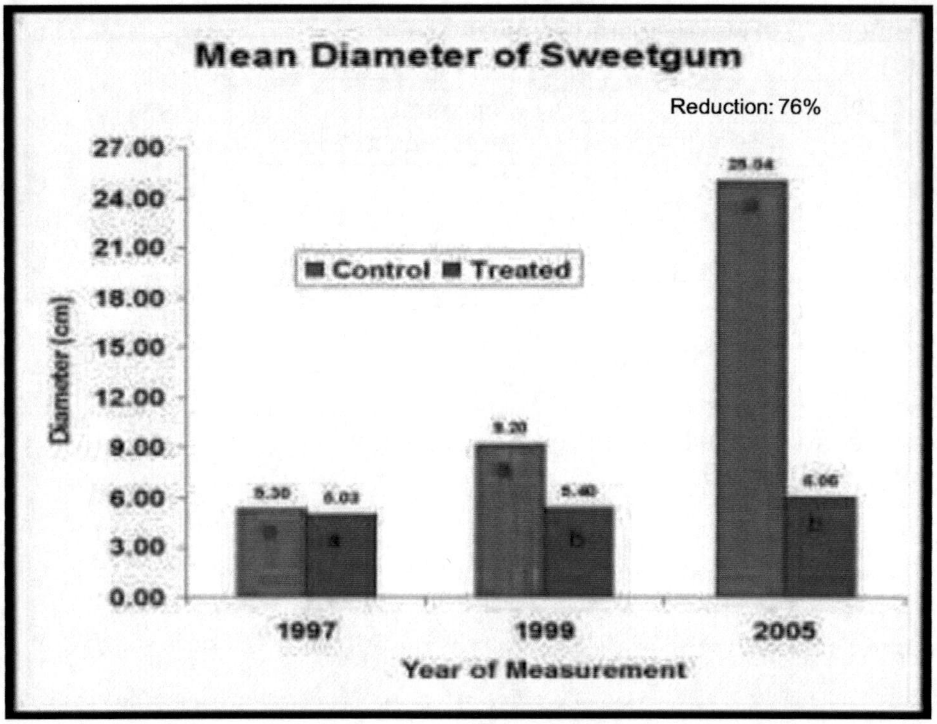

Figure 1. Reduction in diameter growth of sweetgum by PBZ in an eight year experiment. 1997- the year that initiated the PBZ treatment and the measurement was taken before the treatment. 1999 – the measurement was taken at the end of 2 years of the PBZ treatment. 2005 – the measurement was taken at the end of 8 years of the PBZ treatment.

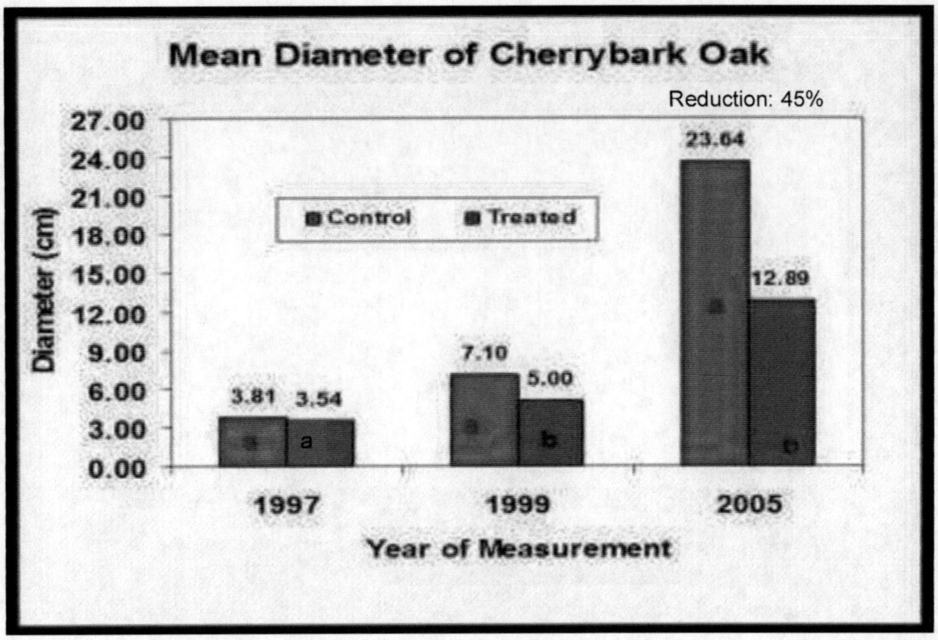

Figure 2. Reduction in diameter growth of cherrybark oak by PBZ in an eight year experiment. 1997- the year that initiated the PBZ treatment and the measurement was taken before the treatment. 1999 – the measurement was taken at the end of 2 years of the PBZ treatment. 2005 – the measurement was taken at the end of 8 years of the PBZ treatment.

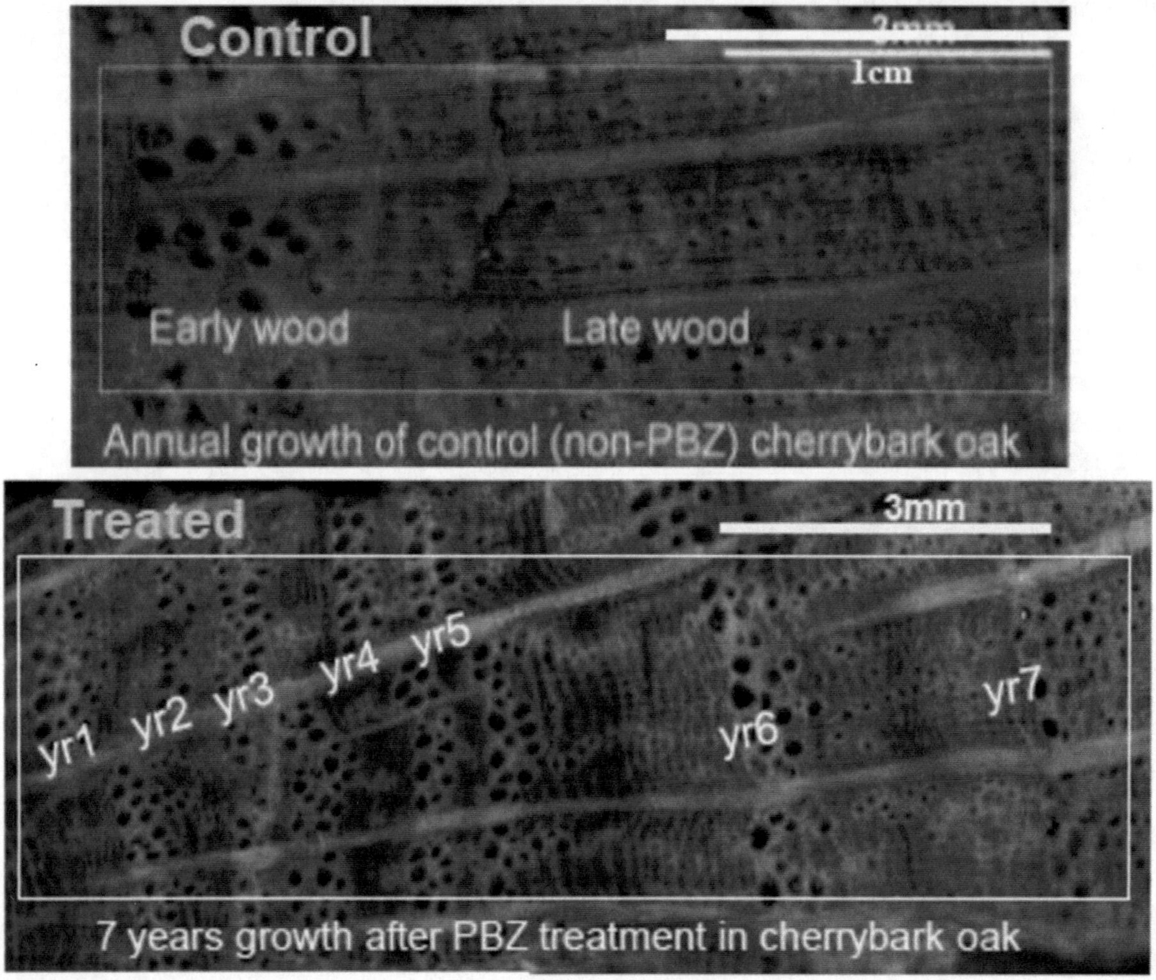

Figure 3. Illustration of the effect of PBZ on cambium growth of cherrybark oak, top pic: showing an annual ring in one year growth of a nontreated tree; bottom: showing a PBZ treated tree's cambium growth across 7 years.